Matthew Cavanaugh

About the Author

ANDREW SULLIVAN is a senior editor at the *Atlantic Monthly* and the American columnist for the *Sunday Times* of London. A frequent television and radio news commentator and former editor of the *New Republic*, he writes daily for www.andrewsullivan.com, one of the first and most popular blogs on the Internet. He lives in Washington, D.C., and Provincetown, Massachusetts.

The
Conservative
Soul

FUNDAMENTALISM, FREEDOM,
AND THE FUTURE OF THE RIGHT

Andrew Sullivan

HARPER ⬤ PERENNIAL

NEW YORK ● LONDON ● TORONTO ● SYDNEY

HARPER ● PERENNIAL

A hardcover edition of this book was published in 2006 by Harper-Collins Publishers.

HarperCollins books may be purchased for educational, business, or sales promotional use. For information please write: Special Markets Department, HarperCollins Publishers, 10 East 53rd Street, New York, NY 10022.

FIRST HARPER PERENNIAL EDITION PUBLISHED 2007.

Designed by Nancy B. Field

The Library of Congress has catalogued the hardcover edition as follows:

Sullivan, Andrew.
    The Conservative soul : how we lost it, how to get it back / Andrew Sullivan—1st ed.
        p. cm.
    ISBN: 978-0-06-018877-1
    ISBN-10: 0-06-018877-4
        1. Conservatism—United States. 2. Fundamentalism—United States. 3. United States—Politics and government—2001– I. Title.
    JC573.2.U6S85   2006
    320.520973—dc22   2006040647

ISBN: 978-0-06-093437-8 (pbk.)
ISBN-10: 0-06-093437-9 (pbk.)

07 08 09 10 11 ID/RRD 10 9 8 7 6 5 4 3 2 1

For Aaron

The true value of a man is not determined by his possession, supposed or real, of Truth, but rather by his sincere exertion to get to the Truth. It is not possession of the Truth, but rather the pursuit of Truth by which he extends his powers and in which his ever-growing perfectibility is to be found. Possession makes one passive, indolent, and proud. If God were to hold all Truth concealed in his right hand, and in his left only the steady and diligent drive for Truth, albeit with the proviso that I would always and forever err in the process, and to offer me the choice, I would with all humility take the left hand, and say: Father, I will take this—the pure Truth is for You alone.

—Gotthold Ephraim Lessing, "Anti-Goeze,"
Eine Duplik (1778)

Contents

Prologue

THIS BOOK WAS BORN OUT OF FRUSTRATION. FOR ALMOST all my life, I've considered myself a conservative. But in the past few years, I've found myself having to explain this more and more. The questions keep mounting up: How can you be gay and conservative? How can you support banning all abortion? How can you have bought into the Iraq war? How can you back dangerous theocrats? How can you support . . . and then you fill in the blank for various politicians whose vacuity and odiousness seem, to the questioner, indisputable.

There have been many times when I have felt like throwing in the towel and simply saying: all right, I'm not a conservative, if that's what it now means. But there have been many other times when I have found myself drawn into long and often interesting arguments about what conservatism can now mean, what it has meant in the past, whether it means the same thing in Britain and America, and whether it now encompasses so many ideas and factions that it can barely be described at all.

This book, for what it's worth, is an attempt to explain what one individual person means by conservatism. It's an attempt to account for what has happened to it to cause such confusion and debate, and why the version I favor is one I still believe is the best way of approaching the exigencies of our current, perilous moment.

My conservative lineage is an idiosyncratic one, and it's worth getting on the table here, just so you can see where I'm coming from. My personal journey may, indeed, make my conservatism idiosyncratic in the current American or British debate. So be it. I cannot change who I am and where I came from. All I can do is make an argument and hope you find something worthwhile in it.

I grew up in Britain in the 1970s and 1980s. I was a teenage Thatcherite, excited by a conservative leader who, by sheer force of will, transformed a country's economy and society from stagnation to new life. In an English high school, I also wore a "Reagan '80" button and saw the former California governor as the West's best hope for survival against socialism and Communism. The conservatism I grew up around was a combination of lower taxes, less government spending, freer trade, freer markets, individual liberty, personal responsibility, and a strong anticommunist foreign policy. I was also a devout Catholic, who felt that my own faith was often scorned or misunderstood by those in power. My faith also told me that there was more to life than politics; and that the best form of politics was that which enabled us to engage in nonpolitical life more fully and more freely.

My intellectual heroes in my teenage years were not unusual for a young right-winger: George Orwell, Aleksandr Solzhenitsyn, Václav Havel, Friedrich Hayek. Yes, I was a nerd, although I somehow managed to avoid a crush on Ayn Rand at any point. But I was also excited by the battle of ideas, and saw the height of ideological combat in the last years of the cold war as an exhilarating time to be alive and thinking. I became interested in politics because I saw how politics could make a difference in the world. And when I lived to see communist tyranny evaporate, and freedom and prosperity spread so widely in the wake of the last great totalitarian nightmare, I felt I had been a witness to something great and ennobling. I still do.

I came to America at the height of the end of the cold war—a few months before Ronald Reagan's reelection, which I heartily backed—and have happily lived in the United States ever since. In the 1990s, when the battle over free markets and communism seemed to have been resolved into a sharp, unexpected victory for the side I had taken, I felt more able to pick and choose in politics, American and British. I had moved to a country well to the right of the one I had left, and found much to like and admire about the place. I still favored Reagan and the first president Bush. But I managed to find the centrist policies of Bill Clinton and Tony Blair congenial in many ways as well—and saw them both as critical to reconciling what was left of the left to the new market economy. I didn't think backing a moderate liberal as some kind of betrayal of conservatism. In fact, I thought it was a sign of conservatism's success that I could do so with so few regrets.

I've never been a partisan, Tory or Republican, because I'm not a joiner by nature. It has never disturbed me that we have two party systems in the Anglo-American West, and I've often felt willing to back a reformed party of the left if the governing party of the right had become exhausted or corrupt. So I endorsed Clinton in 1992 and Blair in 1997. As a conservative, I narrowly backed George Bush over Al Gore in 2000, because I found Gore's newly statist and populist persona to be fake, and worryingly left-wing. By 2004, however, I felt forced to back John Kerry because of the ineptness and nonconservative recklessness I saw in the Bush administration's first term. Some may call this picking and choosing some sort of "flip-flopping." I don't. I consider it an aspect of being an individual, trying to figure out the world as it is, rather than as I might wish it to be.

This book explains how I have come to find myself increasingly estranged from the Republican Party, from the policies it now stands for, and from the philosophy it now represents. My discomfort is shared by many to different degrees, but my own journey can only speak for itself. In saying this, I will inevitably come across as some kind of preening purist, claiming the mantle of "true conservatism" for my own wish-list of ideas, while dismissing others in Republican or Tory ranks as somehow phonies. But that is not my intent. Of course, by favoring one version of conservatism over another, I am not neutral in the argument. But conservatism has become such a large and sprawling complex of ideas that no one has a monopoly on the term anymore. I don't want to get into an emotional and pointless battle over semantics or labels. So let me concede up front that plenty of people who strongly disagree with the analysis and argument of this book are still, in my mind, legitimately be called "conservative."

All I ask in return is an acknowledgment that the kind of politics I favor and argue for in this book is also well within the bounds of the Anglo-American conservative tradition. I may be in a minority these days, but what matters is the argument, not the number of people behind it. Perhaps this conservatism now has a brighter future among Democrats and Independents than among Republicans. But I hope that many disenchanted Republicans—and even more modernizing Tories in Britain—will find plenty to agree with. I also beg forgiveness in advance for the inevitable simplification of very complex events and ideas that are still too close to us for a truly disinterested assessment. This is not an attempt at a definitive account of conservatism. It is an essay in defense of an idea of conservatism now in eclipse.

In that sense, it is doubly conservative. In defending what might seem a lost or losing cause, I have adopted the usual conservative posture of sadness at the pace and direction of current events.

I present in this book two rival forms of conservatism. The first I have called "fundamentalism," and it represents both a form of religious faith and new variety of politics to represent it. The second I call "conservatism," by which I simply mean conservatism as I would describe and explain it. Both have common elements—primarily a suspicion of change and appreciation of the inherited wisdom of the past. But they diverge dramatically in how they see the role of government, how they view faith, and how they understand the intersection of the two. I do not believe this divergence is a trivial or superficial one—in fact, I believe that fundamentalism is, in some respects, the nemesis of conservatism as I have always understood it. I regard its current supremacy not as a continuation of the conservatism of the past, but a usurpation of it.

Rescuing conservatism, I argue, means rejecting the current fundamentalist supremacy in almost every respect. Of course, a Republican politician, trying to lead the current Republican Party, cannot immediately embrace these politics without jeopardizing his chances for power. I am not naive enough to believe that what this book argues for will readily become the Republican or even Tory mainstream. So I understand why this approach may not work today in retail politics. But a conservative writer is luckier than a conservative politician. I have no primaries to win. What follows is simply what I have come to believe—useful or useless, central or idiosyncratic, feasible or out of touch. I hope to persuade you, but if I don't, I aim at least to have helped clarify where we disagree. In such a polarized political climate, that's not such a paltry goal.

The book begins with a brief account of the historical context for the current debate: the end of the cold war, and the triumph of the right in the West. The next two chapters deal with two ascendant conservative approaches to the post–cold war world. Chapter 2 describes the fundamentalist response—the elevation of a set of religious doctrines as the primary means to understand and govern a disillusioned, chaotic world with no secular utopias left. Chapter 3 charts the related bid to resuscitate a philosophy of theologically inspired "natural law" to govern politics, especially cultural politics, after the collapse of the left. Both approaches, I argue, are indispensable to understanding the radically new policies of the Bush administration, domestically and abroad. Only a deep understanding of the fundamentalist psyche and the theoconservative project can explain what has happened to Republicanism in so short a time. In the fourth chapter, I tell the story of how such ideologies created and sustained the legacy of the Bush administration in matters foreign and domestic.

Then, in chapter 5, I turn to my own understanding of what conservatism is: a political philosophy based on doubt, skepticism, disdain for all attempts to remake the world and suspicion of most ambitious bids to make it better. I sketch the contours of such a conservatism: its philosophical modesty, its practical restraint, its radically random notion of history, and its experiential, ritual, and sacramental approach to Christianity. Finally, I describe what such a conservatism says about politics, its scope and dangers, its limits and promise. The book ends with a defense of the conservatism that advances individual liberty and limited government. In that sense, I am not disenchanted. In fact, I think conservatism of the variety I support has a rich and vibrant future ahead of it. It is, I would argue, the only coherent politics for the future of the free West.

I have bitten off a great deal—probably far too much. I've been mulling these thoughts for a couple of decades now—both in the academy as a doctoral student and as a working journalist, reviewer, and writer. I've also mulled them as a person of faith whose attachment to Christianity remains tenacious if not unchastened or uncharged. As any student of the great French essayist Michel de Montaigne would hasten to concede, I also have no doubt that I may change my mind again in the future. It is both alarming and humbling to try and state your beliefs so baldly in one place—and everywhere I look in the text I see further complications and nuances that I want to add or subtract. But there are times when it's helpful to pull your thoughts together, set them down as clearly as you can, draw a line beneath it, and let the readers take the arguments where they want. Think of this book, then, as an opening bid in a conversation, rather than the final summation of a doctrine. This is the best I can do for now. And that, the conservative in me cautions, is enough.

CHAPTER 1

A Silver Age
1989–2001

"The era of big government is over."
—PRESIDENT BILL CLINTON, *January 1996*

I

ALL CONSERVATISM BEGINS WITH LOSS.

If we never knew loss, we would never feel the need to conserve, which is the essence of any conservatism. Our lives, a
series of unconnected moments of experience, would simply
move effortlessly on, leaving the past behind with barely a look
back. But being human, being self-conscious, having memory,
forces us to confront what has gone and what might have been.
And in those moments of confrontation with time, we are all
conservatives. Sure, we all move on. In America, the future is
always more imperative than the past. But the past lingers; and
America, for all its restlessness, or perhaps because of its restlessness, is a deeply conservative place.

The regret you feel in your life at the kindness not done, the
person unthanked, the opportunity missed, the custom unob-
served, is a form of conservatism. The same goes for the lost love
or the missed opportunity: these experiences teach us the fragility
of the moment, and that fragility is what, in part, defines us.

When an old tree is uprooted by a storm, when a favorite

room is redecorated, when an old church is razed, or an old factory turned into lofts, we all sense that something has been lost—if not the actual thing, then the attachments that people, past and present, have forged with it, the web of emotion and loyalty and fondness that makes a person's and a neighborhood's life a coherent story. Human beings live by narrative; and we get saddened when a familiar character disappears from a soap opera; or an acquaintance moves; or an institution becomes un-recognizable from what it once was. These little griefs are what build a conservative temperament. They interrupt our story; and our story is what makes sense of our lives. So we resist the interruption; and when we resist it, we are conservatives.

There is a little conservatism in everyone's soul—even those who proudly call themselves liberals. No one is untouched by loss. We all grow old. We watch ourselves age and decline; we see new generations supplant and outrun us. Every human life is a series of small and large losses—of parents, of youth, of the easy optimism of young adulthood and the uneasy hope of middle age—until you face the ultimate loss, of life itself. There is no avoiding it; and the strength and durability of the conservative temperament is that it starts with this fact, and deals with it. Life is impermanent. Loss is real. Death will come. Nothing can change that—no new dawn for humanity, no technological wonder, no theory or ideology or government. Intrinsic to human experience—what separates us from animals—is the memory of things past, and the fashioning of that memory into a self-conscious identity. So loss imprints itself on our minds and souls and forms us. It is part of what we are.

It is no big surprise, then, that the first great text of Anglo-American conservatism, Edmund Burke's *Reflections on the Revolution in France*, is all about loss. It's a desperate, eloquent, sweeping screed against the wanton destruction of an old order. When the French revolutionaries stormed the Bastille, overthrew a monarchy and a church, remade the calendar, and executed dissidents by the thousands for the sake of a new, blank slate for humanity, Burke felt—first of all—grief. His primary impulse was to mourn what was lost. He mourned it even though it wasn't his. This was not the same as actually defending the old order, which was, in many ways, indefensible, as Burke conceded. It was simply to remind his fellow humans that society is complicated, that its structure develops not by accident but by evolution, that even the most flawed bonds that tie countless individuals are not to be casually severed for the sake of an inchoate idea of perfection.

Even people and societies with deeply wounded pasts, with histories that cry out for renewal and reform, are nonetheless the products of exactly that past. They can never be wiped clean, born again, remade overnight. Even radical change requires a reckoning with the past if it is to graft successfully onto a human endeavor or life. An alcoholic who becomes sober will be unlikely to succeed without a thorough and often painful accounting of how she came to be where she is; and her future will be pivoted against her past and unimaginable without it. When those in recovery insist that they are still alcoholics, they are merely saying that they are human. They have a past. And living in the present cannot mean obliviousness to one's own history. It means living through and beyond that history.

If that is a conservative insight, it presses more powerfully than ever today. If the essence of conservatism is conserving, then our current moment is an extremely unnerving one. In the twenty-first century, the pace of change can at times seem overwhelming. Quantum leaps in technology have transformed how we communicate with one another and expanded everyone's access to an endless array of life possibilities. Jobs last months rather than decades. Travel is cheaper and easier than ever before. Mass immigration has altered settled cultures across the globe. Freer trade has upended life's certainties and customs in almost every society on earth. Assumptions we once made about who we are as a people, or as a country, or even as men and women, are now open to debate. The meaning of family, of marriage, of health, of sex, of faith, are now things we cannot simply take for granted as a shared understanding. And this pace adds a more bewildering and communal sense of loss to the more familiar, quotidian losses of human existence.

Adults face this first of all as they try to bring up their children. They look around them in the twenty-first century and they increasingly see no stable cultural authorities to tell them what virtues to instill in the next generation. They see a media industry geared entirely to profit, impervious to the needs of children, and indifferent to the distinction between high and low culture. They see church hierarchies indelibly tarred by sexual abuse and preachers indistinguishable from politicians. They see authoritative cultural institutions—like the network news, the BBC, or the *New York Times*—nibbled away by a thousand blog-bites or self-inflicted embarrassments and errors. They see no common culture, able to direct and uplift them to a simple, worthy goal. They see only subculture after subculture, each competing with the next. They know that something more durable and valuable must be out there—how did our own parents and grandparents do it?—but they seem less supported and more isolated than ever. This makes even die-hard liberals conservatives after a fashion. And it makes our age a conservative one—for both good and ill.

This disorienting cultural and social change encompasses the trivial and the profound. I was watching MTV the other day and noticed something. On the most popular music video request show, the videos themselves are now edited for a shorter attention span. What was once a three-minute visual blizzard is now a mere thirty seconds' worth of blur. The attention deficit MTV helped create has now cannibalized MTV itself.

I'm a Catholic, baptized into a church that defines itself as immune to change. And yet, in my lifetime, the very language of the Mass has changed, the liturgy has been altered, the old, symbiotic relationship between community and church has been severed, the deference of laity to clergy has waned, and the pews have emptied more quickly than anyone predicted. Partly in response to this, the Catholic hierarchy has attempted to restore and reimpose old certainties by force and fiat, to reassert unquestioned authority, and to reverse or freeze the changes of the last half century. But this too has only intensified either estrangement or division or combat, proving that breaking something is far easier than building something, that our common bonds are more easily wounded than healed.

These are two personal examples of a far bigger picture. The great motors of global transformation—the ferocious economic growth of China and the pitiless cultural churn that is America—are propelling a worldwide sense of dislocation as profound as any in the past. The migrants of central Africa, the outsourced tech workers of India, the construction workers in China, the unemployed of continental Europe, and the technological innovators of America are all flotsam in this global riptide. We are all far away from home now.

Global change is not new, of course. But there are moments in human history when cumulative evolution becomes a sudden exponential leap, when the rapidity of change everywhere, combined with a global awareness of it, feeds on itself. Today's economic and social revolutions—from the mass migrations in Africa and China to the technological revolutions in the nerve centers of the West—catalyze each other. The change that human beings have always experienced is now amplified by a deeper and wider network of actors, and a faster means of communication, leaving the levers of change more distant from us and the sense of helplessness more profound.

Technological breakthroughs in one place are now breakthroughs everywhere. The DNA of the 1918 flu is posted on the Web, and anyone with a modem can look for it. Gay teens in Africa and Kazakhstan now dream, in ways previous generations couldn't, of social liberation in other countries. And their very dreams alter their reality, if only by casting it into shade. Islamic radicals communicate on blogs and Web sites and show atrocities on a URL accessible at the same time in Scotland and Tasmania. Al Jazeera broadcasts anti-Zionist and anti-Western screeds, but simultaneously channels images of free women and open societies to places where women are domestic slaves and freedom of speech is anathema. A flat world has fewer and weaker levees to hold back new floods of social change and innovation. They threaten to drown those few enclaves still untouched by them.

These are bewildering times. This would be true on a purely social and economic plane. But it is also true in the human consciousness. Traditional societies have ceded to far more dynamic ones. Where once many towns and even cities in the West could assume broadly shared cultural and religious values, today that is decreasingly the case. Multireligious societies, multiethnic societies, multiracial schools, multicultural suburbs: these are now the background of our lives. We live in niches and communicate across chasms. The truth was once delivered daily to our doorsteps in a single newspaper or declaimed in the authoritative tones of Walter Cronkite or Alastair Cooke. Now there is a cacophony of bloggers and podcasters and viral videos and cable channels and indie movies and online petitions. Each has its meaning and small piece of authority; and in response, we each have to try to develop our own.

These are bewildering times for the empowered, let alone the powerless. I am fortunate enough to have won the demographic lottery in my own time. Born into a free, prosperous West, I have had every advantage available to a global citizen in an age of dizzying metamorphosis. And yet it is still unsettling. In a mere twenty years of adulthood, I have gone from writing on an electric typewriter with carbon paper to blogging in real time on the World Wide Web. My own economic niche—writing—has gone from relatively few centers of expensive, exclusive print power to an army of blogging self-publishers. Paper has ceded to pixels. Editing has been outsourced to writers. On the Internet, there are no institutions; there are merely pages. And each page is as accessible as any other.

It should be no surprise, then, that a world full of such loss is also a world full of resurgent conservatism. A period of such intense loss and cultural disorientation is a time when the urge to conserve what we have left is most profound. Just as you might grab onto an old tree in a storm, or move a battered old sofa into a new apartment in a new city, so we cling to the familiar in response to the onslaught of the unknown. We therefore need no complicated explanation for why conservatism is ascendant as a political philosophy and cultural force in the West. The flipside of light-speed economic and technological change is a deep sense of social and cultural loss. When the world is this challenging, stability seems more necessary than ever. When change beyond our control overwhelms us, we naturally seek the familiar as a counterbalance. When the exterior world is this unstable, we seek a clearer form of security within. And the more unstable it gets, the more we tend to demand the extreme forms of certainty that can reassure us most completely, and abate our sense of lostness more profoundly. We look for a rock not of sandstone, but of granite.

II

CONSERVATISM IS RESURGENT FOR ANOTHER REASON AS well. It is so glaring that many of us refuse to look it in the face. The second reason for conservatism's ascendancy is the collapse of what might, for want of a better term, be called the "left." Not so long ago, a school of thought existed that posited that human problems could be solved rationally and coherently through a benevolent government and progressive social policies. Poverty could be ended. Inequality could be rectified. The economy could be run by experts to provide permanently full employment and equal opportunity. Colonialism could be ended to usher in a new era of prosperous self-government in the developing world. Religion was moribund or indistinguishable from superstition and doomed for eventual extinction. If the left saw human life in the nineteenth and twentieth centuries as a problem, then it also proffered a solution. That solution varied from the pragmatic progressivism of the Western welfare states to the totalitarian final solution of Stalin's or Mao's communism. But solution it was: a solution to loss, to instability, to suffering, and even to war.

Western politics was, in liberalism's golden age, mere managerialism—for self-evidently or provably noble ends. The idealism that fueled the British postwar welfare state was also supported by a confidence in intellectual circles that rationalism and expertise were the ultimate solutions to all human problems. There were many rooms in this large political mansion, and they varied from the optimism of Lyndon Johnson's Great Society to the extremes of social engineering in Stalin's Russia or Mao's China. But they all posited an inevitable march toward human enlightenment, equality, and progress. If John Maynard Keynes promised a permanent end to unemployment, John Kenneth Galbraith foresaw a permanent end to poverty. History was a matter of inevitable progress to the rational maximization of human well-being.

Today it is difficult to believe that anyone ever believed such things, let alone that some could still believe in them. Of course, a handful of socialist utopians hang in there. But they are a tiny minority. The slow collapse of the European social-democratic model began in the 1970s. In Britain, Margaret Thatcher accelerated the process, and Tony Blair consolidated it. In Europe, the left now represents less an optimistic faith in the future than a nervous, insecure refusal to grapple with globalization, and a romanticization of the recent past.

The total collapse of the Soviet Union was, in retrospect, the coup de grâce. Although Western social democracy never aped the authoritarian collectivism of the Soviets, Communist Russia nevertheless represented the most thoroughgoing attempt to bring about the utopian future once dreamed of by Marx. When it was revealed as a corrupt, poverty-ridden shell of a system, the grand idea of the old left never recovered. It died with the stench of evil floating around the corpse. Just as the example of

Nazi Germany stigmatized a form of statist, authoritarian conservatism forever, so the Soviet Union morally and intellectually wounded the left in irreparable fashion around the world.

Perhaps the greatest example of this came in China, once communism's great hope. The most populous country on earth switched ideological allegiance in the late twentieth century, abandoning the insane practices of Maoist communism for a new, Eastern form of authoritarian capitalism. The model remains highly unstable and its future is unknowable. What is indisputable is that the Chinese elite saw that a communist economic model condemned the country to economic and social stagnation and even collapse. And so a new form of authoritarianism came into being—an alliance between a one-party state and corporate capitalism. It was brutal and repressive, but it delivered economic gains unknown to the Maoists, and managed to avoid the full-scale mass murder of the communist past. It may still fail; but what it isn't is as important as what it is. The utopian vision that had first glinted in Marx's eye in the Reading Room of the British Museum in Victorian London was finally buried in the gleaming new office towers of Shanghai in the first decade of the new millennium. If you want to find communism today, you have to wander the ruined slums of Rangoon and Havana to find it. They speak for themselves.

In America, the democratic left also curdled. In office, the last Democratic president pursued policies of fiscal conservatism, welfare reform, and a reluctant, timid intervention to prevent genocide in Europe. Bill Clinton did not return to the sharply progressive tax regime of the 1970s nor to the faith in large social programs of the 1960s. Subsequently, in opposition to the Bush Republicans, most Democrats opposed any reform of the welfare state, balked at a war to free a foreign people of despotism, and favored more fiscal responsibility than their opponents. They retain shards of idealism, but even their most expansive ideas—like universal healthcare—are couched gingerly and with an eye on the pragmatic center. Faced with the civil rights movement of their own time, the gay rights revolution, the Democrats ran for cover. In fact, they became much more adept at running for cover than running for office.

The academic left, for its part, retreated into a subculture of postmodern discourse, a subculture that, at its worst, opposed basic notions of Western freedom: of speech, of trade, of religion. These remnants of the left were far more interested in exposing the iniquities and prejudices of their conservative enemies than in holding up a coherent, progressive, or egalitarian alternative to the new consensus. They found patriotism disturbing, religion unfathomable when it wasn't contemptible, and tradition a mere chance for deconstruction. If they had absolutes, they were somewhat grim ones: the right of a woman to abort an unborn child, and a profound, reflexive skepticism toward America's moral standing in world affairs. When theocratic terrorists murdered more than three thousand innocent people on 9/11, some of the leading lights of the academic left immediately blamed America. And in blaming America, they merely foreswore any attempt to persuade it.

In its more global forms, the rump left became almost indistinguishable from the reactionary right. Anti-globalization activists demonized corporations that brought cheap and plentiful food to the starving and malnourished; they targeted pharmaceutical companies that innovated drugs to cure and treat previously fatal disease; some even rallied to the defense of the Baathist regime in Iraq, found excuses for the Milosevic regime in Serbia, and dabbled in the thinly veiled anti-Semitism of those who see nothing wrong with the Middle East except the existence of Israel.

There may be a coherent politics somewhere in there; and my point here is not to oppose or support it. My point is merely to show that whatever the current left represents, it isn't the lung-filling liberal confidence of the 1950s and 1960s, the sense that profound social problems were amenable to grand government solutions. Just as the early civil rights movement degenerated into bitter black nationalism, and the left's campaign for homosexual equality degenerated into a screed in favor of "queerness" for its own sake, so the optimistic left degenerated into resentment and alienation from the center of American discourse. The neoliberals, those on the left who still wanted to wield power in America and Britain, were essentially conservatives of a more hopeful, pragmatic, socially tolerant type. Their base shifted toward the old haunts of the old Republicans and Tories—the libertarian counterculture of the West Coast and the Yankee flintishness of New England—or the new, aspiring middle class of Britons. And insofar as these neoliberals espoused the old doctrines of the old left, they lost support.

This led to a remarkable era of calm and prosperity. Globalization led to great social unease but also to fantastic new wealth and growth. In America, the Clinton-Gingrich years were a strange mixture of cultural warfare and underlying social and economic contentment. A less lucky country would not have had the luxury to consume itself with the details of sexual harassment civil suits and overreaching presidential impeachments. But as government trod water, as taxes stayed low, as the president declared the era of big government over, the success was palpable. Crime collapsed; divorce, abortion, and teen pregnancy rates declined; the Internet was created; huge strides were made against cancer and HIV; a long, steady boom raised incomes and created gigantic new mountains of wealth for the few. With the culture devolving into a million subcultures, utopian schemes on the left in abeyance, and the government deadlocked between two parties, America's global dominance seemed assured, if a little empty. I remember having lunch with a friend at the time, and we found ourselves asking how much more good news could come down the pike. My friend predicted doom. I laughed. Nervously.

For in the midst of such a blessed time, a vacuum was emerging. The ruling consensus was a successful, Anglo-Saxon one: pragmatic, market-oriented, with tinkering at the margins. The weekly Bible of the elite was the *Economist* magazine. The difference between the policies of, say, the first President Bush and President Clinton were as marginal as the substantive disagreements between Blairites and reformist Tories in Britain. Yes, there were nuances and factions. But the old right-left divide began to melt away into the narcissism of small differences. It was a Democratic president who signed welfare reform and balanced the budget. It was a Labour prime minister who made the Bank of England independent and retained the pound.

But beneath this surface success lay a structural weakness. This ruling end-of-century consensus failed to do two things. It failed to fill the vacuum left by the collapse of left-wing utopianism and it failed to assuage the profound anxieties produced by the accelerating pace of global change. This was not, of course, a failure on its own terms. The whole point of the ascendant market-based consensus is that it understood that global change was a given, and that wealth and opportunity lay in grasping that change rather than fleeing it. Free trade was the elite consensus in both America and Britain; neither country retreated drastically from the lower taxes and freer labor and capital markets established in the 1980s. The answer to anxiety was, in both cases, education: education to prepare people for the challenge of the global economy. As for utopianism, the left had abandoned it, and the right had long eschewed it. In Britain, Toryism had always been defined as the opposite of utopian social engineering. In America, the conservative movement had been invented to resist the statist intervention of liberal big government. It sought meaning in the private sector, in free association, in private religion and public austerity.

But there was within conservatism, especially in America, another tendency. This tendency saw the weak spot of the new consensus and exploited it, for short-term political advantage and long-term cultural gain. To those who feared the bewildering rapidity of social, economic, and cultural change, this new ideology offered clear, global, unchanging, solid truth. To those who longed for a politics of meaning, a politics that provided the unifying coherence once provided by the old left, the new ideology offered a very similar recipe: a powerful, untrammeled central government dedicated to making the world both secure and new again. Within the new conservative consensus, a new and powerful strain emerged, the inheritor of the old left, but more stable, more resistant to criticism, more connected to the psyches and souls of the people it sought to lift up and save.

Pragmatic, "small-c" conservatism's brief moment of hegemony was over. The age of fundamentalism had arrived.

CHAPTER 2

The Fundamentalist Psyche

> "The less one knows how to command, the more urgently one
> covets someone who commands, who commands severely—
> a god, prince, class, physician, father confessor, dogma, or
> party conscience."
>
> —FRIEDRICH NIETZSCHE, "The Gay Science"

I

THE ESSENTIAL CLAIM OF THE FUNDAMENTALIST IS THAT
he knows the truth.

It's a simple, short phrase, but it would be foolish to underestimate its power in today's unmoored West and developing
world. The fundamentalist doesn't guess or argue or wonder or
question. He doesn't have to. He *knows*. This knowledge may
be the fruit of slow, guided spiritual evolution or, much more
probably, sudden revelation—but, from his point of view, it is
still knowledge. The distinction that others make in the modern world—that there is a difference between what we know
empirically and what we believe normatively—is one the fundamentalist rejects.

And what the fundamentalist knows is true. It isn't a propo-
sition, held provisionally, to be tested by further evidence. It isn't
an argument from which he could be dissuaded by something
we call reason. It isn't something that is ever subject to change:
what is fundamentally true now, by definition, must be true for

all time. For the fundamentalist, there is not a category of things called facts and a separate category called values. The values of the fundamentalist *are* facts. God has revealed them in a book that is inerrant, whether that book is the Bible or the Koran; or he has entrusted them to a hierarchy whose interpretation of scripture and tradition and history and nature is authoritative and even, in some cases, literally infallible.

This truth comes from outside the human being who holds it. It is an external truth, brought to him by a book, a sacred text, a mullah, a pope, a guru, a political visionary, or a religious community with an authoritative pastor. The human being may internalize this truth, and it may come to seem a part of him, but it never emerges from within organically. It is not usually the fruit of a journey, but of a moment on that journey. Saint Paul's sudden conversion on the road to Damascus is a classic fundamentalist moment. The same can be said for those who are "born again" into a new Christian reality. Or even those secular types, who fall for a secular form of fundamentalism, like Marxism or Nazism, or some total explanation of human events that commands submission and total acceptance. Fundamentalism, in the sense I am using it, means adherence to this external system of thought and faith. You could call secular fundamentalisms ideologies. You could call religious fundamentalisms the moment a living faith becomes an ideology, a doctrine. But the pattern of thought is the same.

The truth embraced by the fundamentalist encompasses everything. It is a total truth. It is truth with a capital T. It informs us not merely about the origin of the universe, or what happens to us when we die, but, if we follow its inexorable logic, what we are supposed to do today, in the next half hour, and tomorrow, and the next day and the next. In its more neurotic forms,

it encompasses not merely basic moral virtues but even the mi-
nutiae of personal hygiene, diet, clothing, facial hair. It gives us
clear rules for the living of our lives, tells us what is good and
what is evil, and regulates everything from what we wear to
whom we love and how we reproduce. It is not even something
that a believer can seriously question—because the human being who could question such things is already defined by a truth
that is far greater than he is. There comes a point, as Pascal put
it, when one must abandon human reason to seek the deeper
reason accessible only by faith. Or as the current pope has said,
"Without the light of Christ, the light of reason is not sufficient
to enlighten humanity and the world."

For many, this kind of religious faith seems to be a form of
suffocation. They observe the adherents of fundamentalist faith
and they are baffled by the way in which a text or a pope or a
mullah can determine how a person acts, thinks, or even feels.
Or it seems as if this state of religious being requires some sudden "leap" that is inaccessible to nonbelievers or beyond their
understanding. God made man? An individual who lived two
millennia ago being vivid in a modern person's existence? These
are the more engaged questions. The glibber ones—although
they hint at something just as interesting—are often sharper as
well. How could a fundamentalist Jew really believe that eating
an oyster will render him a moral outcast? How can a Muslim
blame and punish the victims of rape rather than the perpetra-
tors? How can an orthodox Catholic believe as a matter of infallible doctrine that the woman who conceived Jesus did not die
but was physically lifted off the ground and whisked vertically
above the clouds? These practices and beliefs seem odd, even
unhinged, to many nonbelievers. At the very least, they seem to
imply a life half lived, or vicariously lived, a form of oppression

from without or within, an extreme deference to something inherently unknowable as if it were fully known.

But in these questions and bafflement, these nonbelievers miss something central to the fundamentalist experience. That central fact is that, from the point of view of the fundamentalist, this experience, far from being suffocating or encumbering, is a form of complete liberation. The extreme manifestations of observance emanate from a deeper, calmer place where faith frees the troubled mind from the burden of existential fear and everyday trembling. We have a fundamental choice, the fundamentalist says. We can live in a constant state of doubt, facing what is unknown and unknowable with our weak, empirical, unreliable skills; or we can embrace a total explanation that liberates the human person from the ordeal of flawed consciousness into the joy of salvation and stable happiness.

Fundamentalism succeeds, in other words, because it elevates and comforts. It provides a sense of meaning and direction to those lost in a disorienting world. It does this by taking you into another world, immune to the corruptions and compromises of this one. The rigid recourse to texts embraced as literal truth, the injunction to follow the commandments of God before anything else, the subjugation of reason and judgment and even conscience to the dictates of dogma: these can be exhilarating and transformative experiences. They have led human beings to perform extraordinary acts of benevolence; and we would be foolish either to condescend to or to underestimate their appeal.

We have countless stories of how the surrender of self to God leads to the profoundest happiness that human beings can experience. William James, in his remarkable series of lectures,

The Varieties of Religious Experience, provided several autobiographical accounts of this happiness. An alcoholic hits bottom and then experiences and retells the sudden presence of Jesus:

> Never with mortal tongue can I describe that moment. Although up to that moment my soul had been filled with indescribable gloom, I felt the glorious brightness of the noonday sun shine into my heart. I felt I was a free man. Oh, the precious feeling of safety, of freedom, of resting on Jesus! I felt that Christ with all his brightness and power had come into my life; that, indeed, old things had passed away and all things had become new.

James also cites another nineteenth-century unlettered fellow who suddenly experiences a vision of Christ that transforms him entirely:

> It took complete possession of my soul, and I am certain that I desired the Lord, while in the midst of it, not to give me any more happiness, for it seemed as if I could not contain what I had got. My heart seemed as if it would burst, but it did not stop until I felt as if I was unutterably full of the love and grace of God.

There are other types of faith, of course. There is the faith that is never extricated from doubt. There is the faith that is once-born, and never experiences a catharsis or conversion like those above. There is the faith that treats the Bible as a moral fable as well as history and tries to reinterpret its truths in the light of contemporary knowledge, history, science, and insight. There is a faith that draws important distinctions between core beliefs and less vital ones—that picks and chooses between doctrines under the guidance of individual conscience. There is the faith that sees the message of Jesus as a broad indicator of how we should treat others, of what profound holiness requires, not as a literally true account in all respects that includes an elaborate theology that explains everything. There is the faith that observes but loses passion, goes through the motions but fails to come fully alive. There is the dry deism of many of America's founding fathers. There's the cafeteria Christianity of, say, Thomas Jefferson, who composed a new, shortened Gospel that contained only those sayings of Jesus that Jefferson inferred were the real words of the real rabbi. There is the open-minded treatment of scripture of today's Episcopalianism or the socially liberal but doctrinally wayward faith of most lay Catholics. There is the sacramental faith that regards God as present but ultimately unknowable, that looks into the abyss and hopes rather than sees. And there are many, many more varieties.

But what all of these faiths have in common is declining popularity in much of the Western world and almost complete absence in the developing nations. This is not necessarily a sign of their inferiority, of course. Modes of religious experience cannot be judged like products, with market share an indicator of their truth. In some ways, the decline of nonfundamentalist models of faith may be a sign of their challenges. They leave the believer with her own choices, with the burden of questioning various doctrines even as she struggles simultaneously to believe them. They do not often impart the intensity of happiness, of spiritual ecstasy, that occasionally overwhelms and uplifts the born-again.

By demanding less absolute obedience, by giving the believer oxygen to form his or her own conscience independently of authority, these faiths also command less intense loyalty. By

THE CONSERVATIVE SOUL · 29

leaving the individual conscience as the ultimate arbiter, these tamer forms of religion allow for far greater political, intellectual, and ideological diversity among their faithful. They can even encompass theological debate about basic issues—and defend them as ways to understand the universe more completely. In these faiths, rituals and ceremonies can become the glue for a religious community, rather than abstract doctrines imparted from a book and guarded by a pastor. In this kind of faith-with-doubt, the truth is always at some distance from the human mind and soul, even while it still lingers insistently in the folds of human consciousness, and profoundly affects, if never actually dictates, the way a human being lives her life.

Fundamentalism, in contrast, purposefully and relentlessly forces an unalterable, precise, external truth into the center of a person's life, and demands complete obedience to it. Its core is not the individual conscience, but God himself, and the decision of the individual to surrender himself to God entirely as the premise of every action he commits and every decision he makes. When you read the texts of Islamic or Christian or Jewish fundamentalists, you see the references to God in almost every sentence, every thought, every argument. Allowing space for human beings to be at some pallid distance from this truth would, for these people, be as absurd as a parent allowing a child to risk his physical safety.

In fact, it's less absurd. The consequences of a foolish decision to, say, allow a young girl to walk too close to the curb of a busy street are, at worst, a wrecked body and life. The consequence of straying from ultimate truth is eternal damnation—an endless, infinite, fathomless torment no one could rationally wish upon herself. So what others describe as "intolerance" is, from the fundamentalist's point of view, an expression of love. If you would do all you can to prevent or dissuade someone from throwing himself off a cliff in despair, you should surely go to great lengths to prevent a neighbor from consigning himself to the endless, hideous torture of eternal damnation. This spans the various faiths. Fundamentalism is an approach, a mind-set, a psyche—rather than a specific doctrine. The fundamentalist Christian's relationship to what he sees as the literal, unquestionable truth of every word in the Bible is not far off Islamic fundamentalists' even less mediated relationship to the Koran, or, in the secular world, an old-school Communist's attachment to party loyalty and Marxist dogma. And so when Christian fundamentalists take young kids confused about their sexual orientation, seclude them in camps, and try and reprogram them with psychological and spiritual pressure, they call the organization "Love in Action." And here's the important point: they mean it. They're sincere.

The appeal of fundamentalism is not somehow irrational either. The issues we are dealing with here matter. More to the point: they matter more than anything else. What the universe means, where our souls are headed, what our souls are, what morality consists in, what good and evil mean: these are questions that obviously matter more than the choice between a cappuccino and a latte, or between a Democrat and a Republican. To place them at the forefront of a human life is not an irrational choice; it is, in many respects, the only rational one. Even those who pretend that they do not need to deal with these things have usually come to terms with them somehow—even if it is by believing that they somehow don't matter or cannot be resolved or can be deferred while we go shopping, surf the Web, or watch the latest reality show.

The fundamentalist argues that deferring these matters is

a dangerously reckless approach to human existence. And the fundamentalist's own knowledge of the fundamental truth demands a radical reorientation of his own life. What he once valued must be thrown away. Even as he lives among the unfaithful, he must try and sustain his own link to ultimate meaning, to the security that surpasses every earthly security. I recall my own faith as a child. I have never had a conversion experience, although I can say that there have been a handful of moments in my own life when I felt the overpowering presence of Christ. I merely grew up in my own Catholic faith the way a plant grows in a solarium. It was the air I breathed, the light I lived by. As an altar boy, I remember getting up in the bright early dawns of the English summer and walking through country lanes and town streets to the sacristy of the local church. There I would put on the starched, simple white vestments that mercifully came in extra-small sizes, and inhale the incense and starch and cooling marble of the antechamber by the altar. In May, in honor of the Virgin Mary, our little church was festooned with blossoms and flowers; and in the hours I spent in the woods and fields around my own house I felt no sense of discontinuity between nature and the supernatural, between the mysteries of heaven and the pied beauty of the earth. I was happy, but not in the ecstatic way William James describes.

And then: an adolescent rupture. I was sent at the age of eleven to a new school for gifted boys, miles away from my home, losing all my childhood friends. It was not a boarding school, so the hour and a half on public transportation there and back each day meant that for months on end in the winter, I never saw the fields and woods of my childhood in daylight. I felt unmoored, displaced, traumatized. And in that period of insecurity, I turned what had been an effortless faith into a deliberate one. My new school was government-run and therefore Anglican. I became suddenly obsessed with doctrinal difference; I was an eleven-year-old with a keen understanding of transubstantiation. In my writing book, I annotated tiny crosses in the margins to ward off evil. In art class, I refused to draw or paint anything that was not somehow related to the Bible. For confirmation, I took the name of Thomas More, England's most famous Catholic martyr and scourge of heretics. I found myself hearing God's call, but in a new and frightening way. One day, He told me not to walk on the cracks in the sidewalk—this was a test of my faith. And so I found myself missing my bus home because I had had to retrace my steps four or five times to atone for a sinful footing.

Looking back, I can see this phase of my faith life as a temporary and neurotic reaction to a new and bewildering school environment and the loss of everything familiar in my life. The neuroses ebbed as I grew more comfortable. But I remember feeling that without the structure of my faith, without my knowledge of its infallible truth, I might have felt completely overwhelmed. I might have gone under. It was the only rock I had; and so I clung to it with white knuckles of certainty, and took its truths and turned them into inviolable words and strictures and commandments. If I violated the slightest one of them, I could lose everything. I had stumbled onto fundamentalism.

The certainty I felt, moreover, translated into a persistent attempt to witness to what others seemed so unaware of. The morning assemblies at my school occurred three times a week in the Anglican church right next to my school. It was a beautiful old building, centuries old. It had existed from before the Reformation, and it still retained the august and intricate marks of the faith that had been banished from it. When I first entered it, along with hundreds of other boys, I was dumbstruck. The others ran to their places in the pews, or they walked casually, chattering, across the altar. No one genuflected, there was no holy water. I saw instantly that what I understood as church was not shared by my new friends. For them, this was a building. For me, their very attitude was a new demarcation between us. Suddenly, I thought, this is how Protestants behave. Since they do not believe that Jesus is present in the Blessed Sacrament on the altar, why should they genuflect anyway? For that matter, why should I? This place was not a real church. Once it had been, when the pope was in charge of the English church. But at some point in the sixteenth century, it had become an empty place where God used to be. And yet I still genuflected. I did it to show them. I did it to protect myself.

This is the fundamentalist mind-set, a fearful insistence on a faith in all its particulars, and an equally clear sense of those who are saved and those who are not. It is obviously not restricted to Catholics, or to young neurotics as I was. The condescension I felt toward my non-Catholic peers is recognizable among born-again Protestants or orthodox Muslims. The same goes for my small but dogged attempts to insist on my difference, to protect my purity, to insulate myself from challenge. What mattered was not asking questions about my own faith— What could transubstantiation really mean? How could fallible men become infallible popes?—but buttressing every nook and cranny of orthodoxy, preparing myself for theological battle, flaunting and daring the secular and Protestant world with my own Catholic convictions. I didn't seek to convert anyone. That, perhaps, is a deep distinction between Catholic fundamentalists and evangelicals. In fact, I think in retrospect that my fundamentalism was a way to seal myself off from others and even, as my budding sexuality began to unfold in my body, to seal myself off from myself.

But I was aware of the enemy: Protestants, atheists, leftists, cafeteria Catholics, secularists. It's perhaps an indication of how distant I was from the actual message of the Gospels that it never occurred to me to love them. Fear them, yes. Love them? How—when my main energy was in protecting myself from their influence? Other forms of fundamentalism have a different tenor. Among evangelical Protestants, or Mormons, the evangelizing mission is integral to the expression of their faith. Where I insisted on genuflection, on weekly communion, on mastering doctrinal issues when most twelve-year-olds were playing video games, fundamentalist Protestants interpreted their faith as an obligation to convert others. That was their mark of difference from the world—and it was bound up in shoring up their own religious identity as well.

This desire to convert is certainly, I think, more in tune with the spirit of Jesus than my repressed and fearful adherence to doctrine and gesture. There is, after all, a profound internal logic to this process whereby a fundamentalist both changes his own life and then attempts to change the lives of others. If you believe that there is an eternal afterlife and that endless indescribable hell awaits those who disobey God's law, then it requires no huge stretch of imagination to make sure that you not only conform to each diktat but that you also encourage and even insist that others do the same. The logic behind this is impeccable. Sin begets sin. The sin of others can corrupt you as well. There are only two solutions: to withdraw from the world and separate oneself as far as possible from evil; or to attempt to construct a world in which such sin is outlawed and punished and constantly purged—by force if absolutely necessary. It is not crazy to act this way if you believe these things strongly enough. In some ways, it's crazier to believe these things and *not* act this way.

In a world of absolute truth, in matters graver than life and death, there is no room for dissent and no room for theological doubt. Hence the reliance on literal interpretations of texts—because interpretation can lead to error, and error can lead to damnation. In its most extreme forms—Islamist worship of the Koran or fundamentalist Christians' reverence for the literal words of the Bible—the very physical handling of the text can be a matter of doctrinal concern. Hence also the ancient Catholic insistence on absolute church authority. Without infallibility, there can be no guarantee of truth. Without such a guarantee, confusion can lead to hell. And the intensity and certainty of the fundamentalist's faith compels him necessarily to extirpate and be on guard for such confusion.

In this project, literalism is essential. Jesus spoke in parables, metaphors, stories that were not intended to be understood as literally true in every respect, but contained a nonliteral truth. His fundamentalist followers today cannot risk such imprecision—because such things can be interpreted rather than merely accepted. And if they can be interpreted, they can also be misinterpreted. Thus evil creeps in. Jesus's message needs to be tidied up, its ambiguities resolved, the gaps in it filled, and everything out of focus in the Gospels brought relentlessly into literal expression.

It's hard not to recall in this respect one of the key moments of fundamentalist art in recent years. The release of Mel Gibson's *The Passion of the Christ* in 2004 was an extraordinary cultural event. The movie was a jaw-dropping success: it brought millions and millions of fundamentalists back into theaters for its depiction of the last day of Jesus's life. And what was striking about the film as an art form was its abandonment of art. In such matters, what was important was veracity and precision, not interpretation and mystery. And so the movie was a masterwork of explicit, fanatical precision. It emphasized not Jesus's message of love and compassion and the necessity to live faith through good works. It focused with astonishing zeal on Christ's suffering as atonement for all human beings for all eternity. Its goal was to insist upon the centrality of Jesus's self-sacrifice as the only thing necessary for human salvation. Or as one fundamentalist critic explained at the time, "The gulf we place between ourselves and God through sin is bridged only by that intense physical agony Gibson depicts and is taken to task for depicting."

For a fundamentalist, this requires obsessing with almost macabre detail on the suffering Jesus experienced. While the Gospels often skip over the details of the Passion, Gibson homes in on it with sometimes fanatical zeal. The centerpiece of the movie is a scene of explicit, unrelenting sadism. It shows Jesus being flayed alive—slowly, methodically, and with increasing savagery. We first of all witness the use of sticks, then whips, then multiple whips with barbed glass or metal. We see flesh being torn out of a man's body. We see pieces of skin flying through the air. We see Jesus come back for more. We see blood spattering on the torturers' faces. We see muscled thugs exhausted from shredding every inch of this man's body. And then they turn him over and do it all again. It goes on for an unrelenting ten minutes. And then we see his mother wiping up masses and masses of blood. What's noteworthy here is that Gibson goes beyond anything even remotely in the Gospels. And he does so because he is concerned above all to be faithful to the doctrine of the atonement. To allow for Jesus to be merely brutalized,

and to see his decision to give himself up as the culmination of
a doctrine of nonviolence and love, would not be sufficient for a
true fundamentalist. He has to show a level of savagery against
Jesus compatible with the fathomless depth of human sin. And
he has to do so as literally as he possibly can.

Great art allows the viewer space to interpret, to ponder,
and to think. Its meaning is often elusive, and designed to be
so. Fundamentalist art views an elusive meaning as an invitation
to error and sin; and so the movie had to remove any autonomy
from its viewers. Gibson achieved this by relentless, stunning,
unstoppable, graphic violence. It gave the viewer the same artis-
tic leeway as a pornographic movie. Toward the end, unsatisfied
with showing a man flayed alive, nailed gruesomely to a cross,
one eye shut from being smashed in, blood covering his entire
body, Gibson had a large crow perch on the neighboring cross
and peck another man's eyes out. Why? Because the viewer has
to be broken down into submission; there can be no doubt about
the violence of Satan—who is, of course, depicted literally in the
movie. And so all the richness and subtlety and grace of centuries of Christian art is literally hammered into an inarguable,
uncontestable demand that the viewer be emotionally brutalized into the sublime self-surrender of fundamentalist faith.

This is what makes fundamentalism hard. It isn't the difficulty of navigating alone through the intellectual and spiritual
challenges of doubt—that is the challenge of nonfundamentalist
faith. What makes fundamentalism hard is the necessity to embrace complete purity, to affirm utter truth in every particular,
to avoid evil at all times, and to be aware of such evil as a constant in one's life. My pubescent obsession with avoiding cracks
in the sidewalk was shadowed by my sense that evil itself was
tempting me to go my own way. In his invaluable book on Chris-

tian fundamentalism *Stealing Jesus*, Bruce Bawer interviewed a former fundamentalist about his mind-set. The man eloquently described the clarity and struggle of the fundamentalist calling:

> There's *no more gray*. You're separated from this world and at the same time inhabiting an unseen world, in which you're fighting an unseen battle against the unseen Enemy. That's what fundamentalism is about: the Enemy. . . . It's a whole different mentality. You not only think about God all the time, you think about the Devil all the time. Everywhere you go, in every encounter with other people, you ask yourself whether this is of the Devil. He's under every bush.

The omnipresence of the antitruth or anti-Christ applies across all fundamentalisms, religious and secular. This was as true of the secular fundamentalisms of the twentieth century as it is for the resurgent religious fundamentalisms of the twenty-first. If Marx had revealed that all consciousness was a function of economic determinism, and that history was leading to a new and inevitable dawn for humanity, then any resistance to that truth in any quarters could only be explained by sabotage or treason or bourgeois counterrevolutionary behavior. This resistance had to be stamped out, because nothing less than human salvation was at stake.

So the secular fundamentalisms of Nazism and Communism penetrated, and had to penetrate, every cranny of the human mind. Those who represented error had to be destroyed—regularly, aggressively, even preemptively. Purges of the priestly elites—the Politburo or the Gestapo or Mao's murderous thugs—were cyclical and continual, like the constant purging of heretics in the history of the Christian church. At times, as in China's Cultural Revolution, or Stalin's manic purges, or Hitler's final solution, the need to extinguish treachery or deviation became a form of suicidal psychosis. The need for total purity, for the world to exist in fact exactly as it exists in a book, often requires mass murder. And when asked to defend the worst excesses, the fundamentalist can always cite the danger of the Devil. That devil is international Jewry, Christian infidels, papist heresy, secular humanism, bourgeois comfort, Zionism, the "homosexual lifestyle," interracial marriage, and any number of others in different times and places. Take any fundamentalism, and you'll find a Satan somewhere.

We should not be in any doubt that exactly this need to extinguish the living rebukes to Holy Truth lay behind the massacre of September 11, 2001. The theology of al Qaeda—and it is unmistakably a theology—regards the mere existence of non-Muslims as an affront to eternal truth. This affront is made doubly offensive by the fact that the infidels have so clearly prospered, while the faithful have languished. The free and vibrant societies of the West, with their market capitalism and religious pluralism, appear to the latest adherents of Muslim fundamentalist paranoia as symbols of Satan. That these Satanic forces have so much power, that they support impure Muslim regimes in the Middle East and, worst of all, that they help protect the Jewish people in land claimed by Islam: all this is an unforgivable assault on the worldview of profoundly religious people like Osama bin Laden.

Bin Laden and his Wahhabist cobelievers are simply repeating what all previous totalitarian theocrats have done. They struck at the enemy—at the symbols of his power and the center of his might. Think of how the strictest of Wahhabists must see New York City: full of Jews living peacefully with Muslims, of gays with protected freedom, of women able to pursue their own dreams of happiness independently of male permission. It is a veritable Sodom crying out for destruction. And the more beleaguered the Islamist fundamentalist feels, the more violent he gets.

We tend to forget these things now, but the overwhelming weight of Christian history lies in a similar syndrome, a profound hostility to religious pluralism or the toleration of error. The notion that human beings' first recourse must be to their own consciences, and that it's perfectly acceptable to be living among sinners or under a government indifferent to religious truth, are relatively new ones in Christian thought. Even John Locke's famous "Letter Concerning Toleration" drew the line at allowing self-professed Catholics to practice their faith openly in Anglican England. The Catholic Church itself only fully embraced such pluralism in the last half of the twentieth century—after close to two millennia of fundamentalism holding sway, with peaks and troughs along the way. Even now, the era of Benedict XVI has seen a revival of the notion that conscience itself is not to be trusted and that papal authority is the only valid source of religious truth.

And this leads to another profound consequence of fundamentalist faith in modernity: alienation. If you believe in your faith as literally and overpoweringly as fundamentalists do, and you live in a society that is not governed by the same truths, or allows competing truths to do battle in the public square, alienation is a rule. How can a fundamentalist Christian or Muslim not feel profoundly alienated in the contemporary West? The West embraces the equality of men and women. Fundamentalist Christianity and, to a much greater extent, Islam, both insist on the separate roles for men and women and the subordination of wives to husbands. The West's universities allow for unfettered intellectual inquiry, the questioning and even outright rejection of the most significant religious truths. Fundamentalist religion regards such intellectual freedom as an agent of error and therefore evil. Homosexuals, abhorred by both Islamic and Christian fundamentalists, enjoy freedoms and even full civil rights in many parts of the West. We should not be surprised when the most influential fundamentalist in America, James Dobson, believes that this will lead to the Apocalypse. The mere allowance of such evils provokes extreme emotional and psychological distress for people who believe in eternal, nonnegotiable truths as fundamentalists do.

These are just a few of the challenges for fundamentalists living in the twenty-first century. The response of Islamic fundamentalists has been, when all else fails, the complete conflation of mosque and state, and totalitarian imposition of religious dogma—as in the Taliban regime in Afghanistan or the Islamist regime in Saudi Arabia or Iran. The response of Christian fundamentalists in societies which constitutionally protect the rights of others to differing views, is to construct a parallel society that helps insulate the faithful from the contamination of sin, and, in the last three decades, a concerted political effort to reverse legal and constitutional protection for behaviors they regard as iniquitous.

These Christian fundamentalists have therefore created an alternative universe of schools, homes, churches, mega-churches, books, movies, music, and television stations in order to protect themselves from the evil influence of what they deem "secular humanism." Their numbers are such that the media itself in America, though previously dominated by those who were either nonfundamentalist believers or nonbelievers, have had to adjust to their sheer market power. And Christian fundamentalists have also, in cycles of political activity, interspersed by periods of social withdrawal, organized to change laws or resist courts that have violated their own biblical sense of what is right and wrong. In the twentieth century, the great fundamentalist campaigns in America were against the legal distribution of alcohol, against racial integration, against the teaching of evolution, and against legal abortion and homosexual equality.

What's interesting to note here is that the old liberal assumptions about religious fundamentalism have been proven wrong. The first assumption—common in the early twentieth century—was that religious fervor, especially the most fundamentalist variety, would inevitably die out as scientific advance and cultural change modernized the world. In fact, America is now far more observant religiously than it has ever been. One of the landmark, recent studies of religion in American history is Roger Finke and Rodney Stark's sociological analysis, *The Churching of America 1776–1990*. They point out that in prerevolutionary America, only 17 percent of Americans were formally "churched," with the highest degree of religious observance in Massachusetts, where the proportion was still a mere 22 percent. By the mid-nineteenth century, that proportion had risen to over 30 percent. By 1980, it was 62 percent. It is even higher today. The American constitution was well aware of the dangers of religious fundamentalism allied to government power, hence the First Amendment. But the founders, aware of many religious movements in their own time, nonetheless never had to grapple with the sheer power and institutional coordination of fundamentalist Christianity that we see in America today.

To put it another way: the kind of fundamentalism we are now witnessing has plenty of historical forefathers, but it is also quintessentially modern. Islamic fundamentalism as an ideology, as Paul Berman has shown in his book, *Terror and Lib-*

eralism, is derived in part from twentieth-century European fascism, and propelled by a growing sense in the Muslim world that the West and its values (typified by America and Israel) have left the Islamic nation behind, disenfranchised and disrespected. This modern Islamism is not very traditional. The Wahhabists are often brutal toward existing Arab governments, traditional monarchies, or clerical elites. They have abandoned nationalism and socialism in different countries for a transnational, individualist but overwhelmingly totalitarian submission to an extreme form of Islamic fundamentalism. They are often more comfortable on a Web site than in a traditional mosque.

Islamism's much milder counterpart in America, Christian fundamentalism, or "Christianism," has seen its most spectacular growth in response to modernizing trends. The "religious right" was born out of disgust and rage at court-imposed racial integration, abortion rights, and homosexual equality. We are surprised to find out that many of the leading Islamo-fascists of our age, like Osama bin Laden, have had easy access to the West, have seen its achievements and success, and yet still resisted them. What we have yet to fully grasp is that Western success is precisely the engine for bin Laden's fundamentalist ideology. When you have lost so badly in the world of ideas and material wealth, you feel compelled to reassert the purest of opposite doctrines in order to regain some kind of psychological and spiritual balance. Against the sheer power of globalized capitalism and human freedom, only the power of an accessible, omnipotent God can keep you sane. And the extreme degree of danger that the alleged "secularism" of the West poses can even be used to justify almost any action, including, in the extreme case of Islamist suicide bombers or the far rarer Christianist bombers of abortion clinics, the use of violence.

II

THE MORE YOU READ OR ENCOUNTER THE ACTUAL TEXTS
and arguments and discourse of religious fundamentalism of all
varieties, the more you see certain psychological and spiritual
patterns that unite them. In the most extensive academic study
of religious fundamentalism I have come across, a typology is
even advanced. It's called The Fundamentalism Project, and it
was directed and edited by Martin Marty and Scott Appleby,
two highly respected theologians. It runs to eight thousand
pages and I cannot claim to have made it through them. But
their own conclusions about what fundamentalism is are worth
reading. They observe several "family resemblances" between
fundamentalisms of all types—Christian, Muslim, Jewish, and
other sects like Mormonism or the Jehovah's Witnesses.
Fundamentalists view religious idealism as the basis for
personal and communal identity. They understand truth to be
revealed and unified. That revealed and unified truth is intentionally shocking—the more outlandish it may appear the more
central to the truth it probably is. They see themselves as part of
a cosmic struggle, view their political and cultural opponents as
agents of evil, and envy the success of modern materialism. They
are invariably led by men, and the maintenance of strict gender
roles is critical to following God's will. God is also engendered
as a male; and the gender of the saving individual—Jesus or
Mohammed or Joseph Smith—is also significant.

At the end of their exhaustive study, several of the theologians single out some ideological tenets that unite fundamentalisms of all kinds. Fundamentalists, the theologians claim, are
mainly concerned with the erosion of religious faith as the basis
for social order and seek to reverse this separation of church

and state; they endorse some sort of Manichean universe in which good and evil compete for victory; they emphasize inerrant texts or infallible authority as the basis for their beliefs; and they hold to some version of millennialism or messianism. They also demand clear behavioral sacrifice on the part of believers, and regulate conformity.

This sacrifice can be financial, as in tithing, or merely behavioral. Obviously a gay man comfortable with his sexual orientation cannot belong to a fundamentalist faith; the same goes for anyone who has sex outside a monogamous marital bond. But a gay man who decides to sublimate his entire sexual being into the maintenance of a rigid religious orthodoxy is often an ideal fundamentalist. His own chastity is a particularly onerous sacrifice for the sake of truth; and such a sacrifice in turn intensifies commitment to the orthodoxy. The longer he retains this sacrifice, the more insistent he is on its necessity. And so you have the well-documented phenomenon of repressed homosexual men being in the forefront of religious campaigns to suppress homosexual behavior in others. If they can be liberated from their own natures, why shouldn't their peers? And if they occasionally falter and have emotional or sexual contact with other men, then their response is either cognitive dissonance or regular bouts of self-laceration and renewed commitment to the theology they are sacrificing their life for.

In the fundamentalist universe, there is often nothing too small that it cannot be monitored or proscribed. The roles of men and women are clearly marked; clothing is often inspected closely; existing social institutions, like the family, are always preferred over other, experimental ways of life. There is a spectrum here, of course. There is a world of difference in fact and effect between the complete subordination of women among Islamist societies and the far milder subservience required by Christian fundamentalists. But the pattern of thought is identical. And the impermissibility of any fundamental deviation absolute.

Take a very basic issue, like the matter of conscience. For many nonfundamentalist Christians, conscience is the ultimate arbiter of what they believe. In fact, the right to believe only what one's own conscience can assent to was at the root of the Reformation and long defined such denominations as the Baptists. The Catholic hierarchy long resisted such an idea—until the Second Vatican Council, when it was endorsed, along with religious freedom and an acceptance of religious pluralism.

Now compare that view of what conscientious faith is with the fundamentalism now ascendant across the globe. Bruce Bawer notes a 1997 essay by David Klinghoffer, a Jewish intellectual who is a major supporter of fundamentalist Christianity. Klinghoffer dismisses the idea of faith being something ultimately decided by one's own conscience, and not by the dictates of a book or a church authority. Such a use of conscience requires no sacrifice, he argues, and has therefore waned in the religious marketplace. It's "cafeteria" religion, a hodgepodge of doctrines picked purely for self-interest, and ultimately unreliable as a guide to what is true. Fundamentalist churches, which do the opposite and demand that their congregants submit their minds and consciences to an authorized, external doctrine, have won many more converts than the lukewarm pieties of conscience-based faith. Klinghoffer writes:

In a religious system centered on an orthodoxy, the system asks the believer to subscribe to a set of faith principles, deriving what it asserts as the Truth about God and the universe, from which also follow definite standards of conduct. After the believer has accepted these principles and sought to order his life by them, he gets the payoff: the experience of God and his Transcendence.

Self-surrender to authority first; conscience and self-determination second. In the late twentieth century, in one of the more remarkable shifts in religious life in the West in decades, this model of faith found a new ally in an increasingly fundamentalist Vatican. The latter days of the papacy of John Paul II and the enormous influence of Joseph Ratzinger, first as head of the Church's doctrinal watchdog and subsequently as pope himself, all pointed to a subtle revocation of the Second Vatican Council's concession to the claims of individual conscience. In 1991, for example, then-Cardinal Ratzinger delivered an address in Dallas, Texas, titled "Conscience and Truth." He couldn't have been clearer about how an individual conscience is by no means, for him, the final arbiter of morality or truth.

For Ratzinger, there were two models of the human conscience. The first is what he called "self-consciousness of the I," which is susceptible to "pseudo-rational certainty." That's what most people would call a thinking conscience. Then there's the deeper conscience, already created by God, lurking beneath what we mistake as our conscience: "an original memory of the good and true (both are identical) has been implanted in us." This authentic conscience is always in line with what the Church authoritatively teaches. It cannot be otherwise: "Man can see the truth of God from the fact of his creaturehood. Not to see it is guilt. It is not seen because man does not want to see it. The 'no' of the will which hinders recognition is guilt." This is the fundamentalist pope's description of what Marxists called "false consciousness." And like dissidents in secular totalitarian systems of thought, the individual conscience is often prone to it. Even when you think you are taking a principled, intelligent moral stand, if you are in disagreement with the pope (or dialectical materialism or the words of the Prophet), you are not in fact exercising conscience. That's a delusion fostered by evil. You are merely demonstrating sin and guilt. There is no conscience distinct from truth, Ratzinger insisted; and since truth has not only been revealed, but is helpfully clarified and explained at all times by the pope and the Church hierarchy, all protestations of "conscience" against church teaching are just further manifestations of sin. The circle is complete here. And it is airtight.

When pressed to defend this view of truth and conscience, fundamentalists argue that there is no alternative. If truth were left up to the individual conscience, then we would live in moral anarchy, with anyone picking and choosing what to believe, junking that sentence from the Bible and retaining another—until nihilism and relativism rule. I wonder what Thomas Jefferson would have said. He did not consider himself a "secular humanist." He was a believing Christian; and yet he used his own mind and soul to winnow the Gospels to what he thought were the essentials.

Jefferson also recognized something that fundamentalists either avoid or deny. A believing Christian logically cannot believe that everything in the Bible is literally true for the simple reason that the Bible is full of factual contradiction. In demonstrating this, it's hard to beat Spencer Tracy's tour de force as Clarence Darrow in the movie *Inherit the Wind.* The film recounts the story of the Scopes trial in Tennessee, where a high school teacher was prosecuted for mentioning evolution. Darrow famously cross-examined William Jennings Bryan, a fundamentalist of the old school, about the contents of the Bible.

The movie is not historically accurate but it's a very bracing elucidation of the fundamentalist dilemma, and the questions put into the mouth of Darrow remain unanswerable ones.

Here are a few. If the earth began six thousand years ago with Adam and Eve, who had only two sons, Cain and Abel, where did Cain's wife subsequently come from? How could Joshua have made the sun stand still if the earth revolves around the sun and not the other way around? If Mary the mother of God remained a virgin even after the birth of Christ, why do the Gospels mention Jesus's brothers? The factual discrepancies between the Gospels alone logically require that they not be read literally. To which the fundamentalist is required to answer: then logic be damned. As soon as he concedes that some story may be a metaphor, or that the men who wrote the Bible at various times in history were capable of error, political motivation, social pressures, culturally specific modes of thought that simply could not be embraced today, then he is lost. For the fundamentalist, it is always all or nothing.

For the nonfundamentalist, life is considerably more fraught. For such a person, the exercise of conscience can be an exacting affair. A nonfundamentalist Christian not only has to try to understand the faith as it is presented in the Bible or by church authorities. He also must make an effort to ask himself constantly whether his thoughts and prayers are rationalizing his own desires rather than seeking the truth itself. The proper exercise of conscience is grueling. It constantly asks whether we are self-deluded or engaging in wishful thinking; and it is always open to further argument or revelation. In my own, long engagement with my Catholic faith, I have had many such moments of struggle. What can three gods in one really mean? Can I really believe that a human being was indistinguishable from

God Almighty? Was the resurrection real or metaphorical? Is the communion wafer I just consumed literally the flesh and blood of a man who died over two millennia ago?

Reaching the answer to these questions—and asking them again and again and again—is not an easy process. There have been periods when I have felt the truth of my faith as powerfully as I have felt a warm current in a cool bay, or the stifling heat of a Washington August. And there have been periods when it has seemed utterly empty, drained, arid, and without passion. There have been times when I have read something that has helped click a certain truth into place; moments when I have found myself opening the Gospels in desperation only to find a passage that spoke directly to me, there and then, with a transcendent voice I recognized as divine. And there have been moments when my reason and my anger have rebelled against diktats that make no sense and rejected a clerical hierarchy immune to engagement. But all these moments amount to a faith-process, not a single line over which salvation is assured and before which damnation is inevitable.

For the fundamentalist, in contrast, there is one moment of real conscience, the moment when he makes the decision to conform his mind and will to an external authority. After that, his sole task is obedience, or, at best, being the best student in a class where there is only one set of right answers, prescribed beforehand (and you're allowed, in fact compelled, to see the answers in advance). Being saved requires constant self-monitoring to ensure that you are not deviating from the correct way of living; but it certainly doesn't require the use of intellectual faculties or spiritual meditation to question the doctrines themselves. Once you're in, doubt disappears. Here's a very clear description of the experience:

My faith frees me. . . . Frees me to make decisions others
might not like. Frees me to do the right thing, even though it
will not poll well. Frees me to enjoy life and not worry about
what comes next.

Those words are from the most powerful Christian fundamentalist in the world, President George W. Bush. He wrote
them in his book, *A Charge to Keep,* published back in 1999. By
his own account, Bush was converted in one moment, and, after
that, his life was different. He took a leap of faith and everything altered. It was liberating, gave him a purpose, and helped,
as we can see from his 1999 remark, provide internal calm in
the midst of grave decisions and certainty about the course he
was on. Any further inquiries into his heart and conscience are
futile, since we cannot know another person as he confronts his
God. But Bush's fundamentalist faith is not atypical.

Nevertheless even the most liberated of fundamentalists
must come to terms with the brokenness of the world around
him. He holds an ideal view of the world in which everything
is resolved, in which everything has a place and a meaning. He
has a vision of heaven; and yet he lives on earth. An Islamic
fundamentalist believes that there is only one God and he is
Allah; he believes that this doctrine must necessarily control
the entire earth. And yet it doesn't. And the power of the West
and existence of Israel operate as daily, unbearable assaults on
this necessary truth. A Christian fundamentalist believes that
Christ's atonement on the Cross for all the sins of humanity
means a new dispensation for humanity, a new day, a vision of
perfection the achievement of which requires only the will to
submit. And yet many of his fellow Americans go blithely on
as if nothing has happened. Worse, they even seek to neutralize

the public expression of this eternal truth, to strip the public square of the ultimate meaning of Christian redemption.

III

HOW TO RECONCILE THIS INCONGRUITY? I HAVE MEN-tioned alienation, withdrawal, or bloody lashing out at heretics. But the other, more familiar recourse is to reassure the believer that today's imperfection will eventually become perfection one day in the future. For many Christians, this means merely wait-ing for death and the afterlife. But for Christian fundamentalists, there is another meaning as well: the fact of a future catharsis, the End-Times when history will end and all doubt will cease. The Apocalypse is the answer, the moment when modernity's contradictions are erased in a new era of Heaven on Earth.

This apocalyptic vision is integral to so many fundamen-talisms. The Nazis believed that a new world of racial purity would usher in a new age of human greatness; and a new human being, the Übermensch. The Communists were assured of a fu-ture moment when the Marxist nirvana would dawn, when all alienation would end, when class conflict would disappear and a new age of freedom dawn: a world where every man could be a philosopher in the morning, a gardener in the afternoon, and a poet in the evenings. Marx got his eschatology from Hegel, but the deeper roots are in the Jewish faith, and the ancient Jewish-Christian squabble about whether the Messiah had or had not come. Today, all three monotheisms contain their own Escha-ton. The Jewish prediction of the Messiah keeps some orthodox Jews in a state of febrile expectation and territorial expansion. Shiite Islam has appropriated and refined many of the formerly

Jewish and then Christian versions of the End of Days. Under Cromwell in fundamentalist-run seventeenth-century England, one of the key groups that flourished was called the Fifth Monarchy. Why fifth? Because the four kingdoms predicted in the Book of Daniel were about to be replaced by the fifth, millennial kingdom: the rule of the saints. Christianity subsequently spawned an entire subreligion of eschatological depth and precision, a nineteenth-century doctrine called "dispensationalism" which lays out the exact nature of the looming Apocalypse in all its gory detail.

This much we can say with certainty: no fundamentalism lacks an apocalypse. And the more extreme the fundamentalism, the closer it gets to the apocalyptic obsession. The best study of these forces in modernity I have read is Gershom Gorenberg's prophetic book, *The End of Days*. Gorenberg notices bin Laden's confidence in his 1998 statement: "We are sure of Allah's victory and our victory against the Americans and the Jews, as promised by the Prophet, peace be upon him: Judgement Day shall not come until the Muslims fight the Jews, where the Jews will hide behind trees and stones, and the tree and the stone will speak and say, 'Muslim, behind me is a Jew. Come and kill him.'" For Osama, as with the evangelical Christian right, there was a perfect Edenic past, a fallen present, and a perfect future promised. The heyday of Muslim expansion was also a period when millennialism and a belief that the end of the world was imminent was widespread among Muslims. That belief is also gaining ground fast today. The new belief is more tinged with fanatical anti-Semitism than in the past, but its meaning is still clear. There will be a period of moral degeneracy, followed by the emergence of a false messiah, al-Masih al-Dajjal, who will be a Jew. Jesus—yes, Jesus—will then return, and, in many versions, be joined by a figure called the Mahdi. The two of them defeat the anti-Christ and install universal Islam.

This particular brand of the faith is predictably strong among the most extreme of the Islamists. The *Daily Telegraph* reported in 2006 that Iran's president, Mahmoud Ahmadinejad, is a strong millennialist, as is his entire government elite:

> A common rumor—denied by the government but widely believed—is that Mr. Ahmadinejad and his cabinet have signed a "contract" pledging themselves to work for the return of the Mahdi and sent it to Jamkaran. Iran's dominant "Twelver" sect believes this will be Mohammed ibn Hasan, regarded as the 12th Imam, or righteous descendant of the Prophet Mohammad. He is said to have gone into "occlusion" in the ninth century, at the age of five. His return will be preceded by cosmic chaos, war and bloodshed. After a cataclysmic confrontation with evil and darkness, the Mahdi will lead the world to an era of universal peace . . . Mr Ahmadinejad appears to believe that these events are close at hand and that ordinary mortals can influence the divine timetable.

In this, Ahmadinejad is being orthodox. And it was no accident that a defining moment in Iraq's post-totalitarian chaos came when Jihadists bombed a mosque dedicated to the twelfth imam, in Samarra. In America, this view of the end of the world is extremely common among fundamentalists, and one of the sources of their astonishing growth in popularity. The best-selling series of adult books in the United States is the Left Behind series, which gives a fictional rendering of the Christian fundamentalist version of Ahmadinejad's vision. These books also portray an anti-Christ—a Romanian in charge of the United Nations—although other fundamentalists, such as Jerry

Falwell, insist that the anti-Christ has to be Jewish. In the End-Times, after faithful fundamentalists are brought into heaven in an astonishing Rapture, the rest of humanity struggles through a hideous period of Tribulation, with nuclear war, famine, and plague. In this period, the anti-Christ persuades Israel to build the Third Temple, so he can subsequently destroy it. When this happens, Christ's reign can truly begin, and after he destroys the anti-Christ, he focuses on every nonbeliever, including all the unconverted Jews. The last Holocaust will be followed by the next one—this time with Jesus, not Hitler, at the helm.

Judaism has a similar fundamentalist strain: the notion that the Messiah will come only when the land of Israel has been united again, and when purified Jews can reenter the rebuilt Temple. This strand has become far stronger in modernity, and was given real impetus by the establishment of the state of Israel in 1948 and the capture of all of Jerusalem itself in 1967. To some fundamentalist Jews, all this portends the building of the Third Temple and the arrival of the Messiah. Biblical promises of all of Judea and Samaria fueled the settler movement on the West Bank, and galvanized the radical Jewish elements that murdered prime minister Yitzhak Rabin for offering sacred land for peace and refused to coexist with Palestinian Arabs. In this, of course, they have a strange alliance with the American fundamentalist movement, which is also convinced (for rather different reasons) that the expansion of Israel is a portent of Armageddon. In 2006 one of the most popular televangelists in America, Pat Robertson, said that Israeli prime minister Ariel Sharon had suffered a stroke because he had flouted God's will by ceding some Israeli land to the Palestinians. He subsequently apologized for the timing of his remarks, but not for the theology that propelled them. How could he? For fundamentalists, the Apocalypse is real, and thwarting it or impeding it is a mark of Satan. In retrospect, Sharon got off lightly.

Most people in the West do not want to believe that religious fundamentalism is the prime force behind the wave of Islamist terror that plagued the world in the new millennium. They want to believe that this psychotic violence is somehow a function of a distortion of fundamentalist faith, rather than its logical, ultimate, extreme consequence. Similarly, most secular Americans refuse to believe that Pat Robertson represents anything but a lunatic fringe when he speaks of God's direct involvement in human affairs in order to precipitate the end of the world. But that is a function of not taking fundamentalism seriously enough. It's a function of misreading plain English.

The Left Behind series sells millions and millions of copies. Its key creator, Tim LaHaye, is an immensely well-connected figure, close to the Republican Party, whose wife runs the powerful fundamentalist lobby, Concerned Women for America. As Gorenberg notes, John Hagee's *Beginning of the End: The Assassination of Yitzhak Rabin and the Coming Anti-Christ* topped American Christian best-seller lists in 1996. Every major Christian right figure has invoked the apocalypse at some point or other, whether it was Jerry Falwell's prediction in 1999 that the anti-Christ would be a Jewish male, or James Dobson's 2005 prediction of the "destruction of the earth" prompted by the legalization of same-sex marriage in Massachusetts.

The spokesmen for such a movement are quite candid about their goals and motivations. And because the End-Times usher in a new era of unprecedented horror, many true believers also think it liberates them to do things otherwise proscribed by their own faiths. Gorenberg notes how Jewish millennialists of two millennia ago (they got their timing wrong) were known

THE CONSERVATIVE SOUL · 57

as Zealots and crossed clear moral boundaries by assassinating Jewish priests on holy days or betraying allies. He quotes the political scientist David Rapoport: "Terror is attractive in itself to messianists just because it is outside the normal range of violence and for this reason represents a break with the past, epitomizing the . . . complete liberation which is the essence of messianic expectation."

There's that word again: liberation. When we re-create in our heads the mind-sets of those men who hijacked airplanes on a beautiful September morning, we recoil from thinking of them as embracing freedom. But in their minds, they were. They were part of a broader plan, a reaction to the sacrilege of infidels on Islamic soil, and a precipitant of a wider and wider conflict, leading to a new Caliphate and the apocalypse itself.

When asked to sum up his message to the people of the West, Osama bin Laden couldn't have been clearer. "Our call is the call of Islam that was revealed to Muhammad," he said. "It is a call to all mankind. We have been entrusted with good cause to follow in the footsteps of the Messenger and to communicate his message to all nations." His is not a shrewd political campaign. It is a religious war against "unbelief and unbelievers," in bin Laden's words. At the Hajj in 2006, a leading mullah gave the following speech at Mecca:

> Today, the impure world Zionism, in the modern Age of ignorance, has emerged with the same (polytheistic) ideology, but with new methods. It wants to take over the fields of economy, of culture, and politics, as well as the military, throughout the world. . . . We, the pilgrims who have come to the house of God, condemn the plots and the measures taken by the international Zionism, the deceitful Satan who spreads heresy, polytheism, and idolatry, enslaving human

beings with a new method. It abuses the divine religion of Moses. It takes Satanic measures, and arouses the world's hatred towards this divine religion, and its true followers.

Are these cynical words designed merely to use Islam for nefarious ends? We cannot know the precise motives of bin Laden or other sheikhs and religious leaders, but we can know that they would not use these words if they did not think they had salience among the people they wish to inspire and provoke. This form of Islam is not restricted to bin Laden alone. Its roots lie in an extreme and violent strain of Islam that emerged in the eighteenth century in opposition to what was seen by some Muslims as Ottoman decadence but has gained greater strength in the twentieth, as it drew on Western models of secular totalitarianism. For the past two decades, this form of Islamic fundamentalism has wracked the Middle East. It has targeted almost every regime in the region and, as it failed to make progress, has extended its hostility into the West. From the assassination of Anwar Sadat to the fatwa against Salman Rushdie to the decade-long campaign of bin Laden to the destruction of ancient Buddhist statues by the Taliban to the World Trade Center massacre and the murder of Dutch filmmaker Theo van Gogh there is a single line. That line is a fundamentalist, religious one. And there is no point denying it.

Islamist fundamentalism, moreover, does not restrict its ire to nonbelievers. Like all fundamentalist religions, it is often just as concerned, if not more concerned, with Muslim apostates or heretics. And so, in the murderous maelstrom of post-Saddam Iraq, the foot soldiers of the new religious orthodoxy primarily targeted other Muslims—for collaboration with infidels, for theological errancy, or for mere passivity in the face of "crusader" aggression.

This use of religion for extreme repression, and even terror, is not of course restricted to Islam. For most of its history, Christianity has had a worse record. From the Crusades to the Inquisition to the bloody religious wars of the sixteenth and seventeenth centuries, Europe saw far more blood spilled for religion's sake than the Muslim world did. And given how expressly nonviolent the teachings of the Gospels are, the extremism of Christian fundamentalism in this respect was arguably more striking than bin Laden's fundamentalist version of Islam. But it is there nonetheless.

In this vision of Christianity, violence is very prominent. Here is a description from the Left Behind novels of what will happen in the time of Tribulation: "Men and women soldiers and horses seemed to explode where they stood. It was as if the very words of the Lord had superheated their blood, causing it to burst through their veins and skin. . . . Even as they struggled, their own flesh dissolved, their eyes melted and their tongues disintegrated." Remember: it's Jesus behind the slaughter. The fundamentalist God is a violent one, punishing, terrifying, authoritarian. Those who differ from fundamentalists are often depicted as violent oppressors demanding a violent response, if only in self-defense. "We must defend the family from the assault of hell," declares James Dobson. "Sickness and death befall those who play Russian roulette with God's moral imperatives," he later added. Rallies for Promise-Keepers, an all-male fundamentalist organization, are full of military rhetoric. Bawer cites one typical call at one rally: 'The fiercest fighting is just ahead . . . Let's proceed. It's wartime!" According to the *Nation* magazine, one Promise-Keepers director has called his group an "army of God" that he views as "the fulfillment of the Bible's prophecy of a great force that will destroy sinners and infidels in the period preceding Armageddon."

But put aside the violent rhetoric and extremist logic, and view the universe as perceived by the fundamentalist. For them, the violence and the extreme dichotomy between good and evil make complete sense. Once you have accepted the premises of a total surrender to God's will and plan, what most people view as pragmatic, self-interested, rational activity is anything but pragmatic, self-interested, or rational. It's a capitulation to evil, a compromise with forces that brook no compromise, a pragmatism that merely gives the enemy time and space to strike back, with all the power that Satan possesses. Remember: *there can be no gray.*

IV

DOSTOYEVSKY'S GRAND INQUISITOR MAKES THE LOGICAL case for fundamentalism perhaps as well as anyone. In the story told by Ivan Karamazov in *The Brothers Karamazov*, Jesus returns to earth during the Spanish Inquisition. On a day when hundreds have been burned at the stake for heresy, Jesus performs miracles. Alarmed, the Inquisitor arrests Jesus and imprisons him with the intent of burning him at the stake as well. What follows is a conversation between the Inquisitor and Jesus. Except it isn't a conversation because Jesus says nothing. It is really a dialogue between two modes of religion, an exploration of the tension between the extraordinary, transcendent claims of religion and human beings' inability to live up to them, or even fully believe them.

According to the Inquisitor, Jesus's crime was revealing that salvation was possible but still allowing humans the freedom to refuse it. And this, to the Inquisitor, was a form of cruelty.

When the truth involves the most important things imaginable—the meaning of life, the fate of one's eternal soul, the difference between good and evil—it is not enough to premise it on the capacity of human choice, or to live in a world where differing and competing versions of the truth are allowed to flourish. That is too great a burden for weak humanity. Choice leads to unbelief or distraction or negligence or despair. What human beings really need is the certainty of truth, and they need to see it reflected in everything around them—in the cultures in which they live, enveloping them in a seamless fabric of faith that helps them resist the terror of choice and the abyss of unbelief. This need is what the Inquisitor calls the "fundamental secret of human nature." He explains:

> These pitiful creatures are concerned not only to find what one or the other can worship, but to find something that all would believe in and worship; what is essential is that all may be together in it. This craving for community of worship is the chief misery of every man individually and of all humanity since the beginning of time.

This is the quintessential voice of fundamentalism. Faith cannot exist alone in a single person. Indeed, faith needs others for it to survive—and the more complete the culture of faith, the wider it is, and the more total its infiltration of the world, the better. It is hard for us to wrap our minds around this today, but it is quite clear from the accounts of the Inquisition and, indeed, of the religious wars that continued to rage in Europe for nearly three centuries, that many of the fanatics who burned human beings at the stake were acting out of what they genuinely thought were the best interests of the victims.

With the power of the state, they used fire, as opposed to simple execution, because it was thought to be spiritually cleansing. A few minutes of hideous torture on earth were deemed a small price to pay for helping such souls avoid eternal torture in the afterlife. Moreover, the example of such government-sponsored executions helped create a culture in which certain truths were reinforced and in which it was easier for weak people to find faith. The burden of this duty to uphold the faith lay on the men required to torture, persecute, and murder the unfaithful. And many of them believed, as no doubt some Islamic fundamentalists believe today, that they were acting out of mercy and godliness.

This is the authentic voice of the Taliban. It also finds itself replicated in secular form. What, after all, were the totalitarian societies of Nazi Germany or Soviet Russia if not an exact replica of this kind of fusion of politics and ultimate meaning? Under Lenin's and Stalin's rules, the imminence of salvation through revolutionary consciousness was in perpetual danger of being undermined by those too weak to have faith—the bourgeoisie or the kulaks or the intellectuals. So they had to be liquidated or purged. Similarly, it is easy for us to dismiss the Nazis as evil, as they surely were. It is harder for us to understand that in some twisted fashion, they truly believed that they were creating a new dawn for humanity, a place where all the doubts that freedom brings could be dispelled in a rapture of racial purity and national destiny.

It is even harder for us to believe that they were actually afraid of the Jews. Their own conspiratorial pathology made the Jews out to be almost superhuman in their power and skills. If any were left after the Holocaust was begun, they would wreak a terrible revenge on the allegedly beleaguered German people. So they had to be incinerated en masse—before they could regroup and destroy every last Aryan on earth. You see this paranoia—although in much milder forms—in contemporary fundamentalist Christianity. A leaflet distributed by the Republican National Committee in the 2004 election claimed that a vote for the Democrats would mean that the Bible would be banned in America. One of the dominant themes of contemporary American fundamentalism is that the country is on the brink of "criminalizing" the practice of Christianity. A major conference was held in Washington, D.C., in the spring of 2006, dedicated to resisting the "War Against Christians."

America, of course, is not a neophyte in witnessing the power and appeal of fundamentalism or paranoia. The United States has seen several waves of religious fervor both before and since its founding. The first European Americans were fundamentalists, determined to pursue their vision of religious orthodoxy in a world where their enemies couldn't reach them. Their ships arrived as laden with inerrant scripture as with opaque notions about human freedom. America is a country that has, at various moments, seen slavery endorsed as God's will and alcohol made illegal.

But until recently American evangelicalism tended to keep some distance from governmental power. The Christian separation between what is God's and what is Caesar's—drawn from the Gospels—helped restrain the inexorable theological logic of fundamentalism in America for a long time. The last few decades have proved an exception, however. As modernity advanced, and the certitudes of fundamentalist faith seemed mocked by an increasingly liberal society, evangelicals mobilized and entered politics. Their faith and zeal sharpened, the temptation to fuse political and religious authority beckoned more insistently.

The result is today's Republican Party, which is perhaps the first fundamentally religious political party in American history.

Mercifully, violence has not been a significant feature of this trend in America—but it has not been completely absent. The murders of abortion providers show what such zeal can lead to. And indeed, if people truly believe that abortion is the same as mass murder, then you can see the awful logic of the terrorism it has spawned. This is the same logic as bin Laden's. If faith is that strong, and it dictates a choice between action or eternal damnation, then violence can easily be justified. In retrospect, we should be amazed not that violence has occurred—but that it hasn't occurred more often.

The critical link between Western and Middle Eastern fundamentalism is surely the pace of social change. If you take your beliefs from books written more than a thousand years ago, and if you believe in these texts literally, then the appearance of the modern world must truly terrify. If you believe that women should be consigned to polygamous, concealed servitude, then Manhattan must appear like Gomorrah. If you believe that homosexuality is a crime punishable by death, as both fundamentalist Islam and the Bible dictate, then a world of same-sex marriage is surely Sodom. It is not a big step to argue that such centers of evil should be destroyed or undermined, as bin Laden does, or to believe that their destruction is somehow a consequence of their sin, as Jerry Falwell argued. Look again at Falwell's now infamous words in the wake of September 11: "I really believe that the pagans, and the abortionists, and the feminists, and the gays and lesbians who are actively trying to make that an alternative lifestyle, the A.C.L.U., People for the American Way—all of them who have tried to secularize America—I point the finger in their face and say, 'You helped this happen.'"

And why wouldn't he believe that? He has subsequently apologized for the insensitivity of the remark but not for its theological underpinning. He cannot repudiate the theology—because it is the essence of what he believes in and *must* believe in for his faith to remain alive.

The other critical aspect of this kind of faith is insecurity. American fundamentalists fear they are losing the culture war. They are terrified of failure and of the Godless world they believe is about to engulf or crush them. They speak and think defensively. They talk about renewal, but in their private discourse they expect damnation for an America that has lost sight of the fundamentalist notion of God. They speak of their own imminent persecution and leap immediately to the idea that Christianity itself will soon be criminalized. James Dobson, the most powerful and influential of America's evangelical leaders sees his faith as under direct attack from American government and especially the courts charged with interpreting the Constitution. "I heard a minister the other day talking about the great injustice and evil of the men in white robes, the Ku Klux Klan, that roamed the country in the South and they did great wrong to civil rights to and to morality," Dobson remarked in 2005. "And now we have black-robed men," he added, in reference to the Supreme Court. These are not the words of a confident man. They are the words of someone who feels under intense and unrelenting attack.

Similarly, Muslims know that the era of Islam's imperial triumph has long since gone. For many centuries, the civilization of Islam was the center of the world. It eclipsed Europe in the Dark Ages, fostered great learning, and expanded territorially well into Europe and Asia. But it has all been downhill from there. From the collapse of the Ottoman Empire onward, it has been on the losing side of history. The response to this has been an intermittent flirtation with Westernization but far more emphatically a reaffirmation of the most irredentist and extreme forms of the culture under threat. Hence the odd phenomenon of Islamic extremism gathering steam so rapidly in the last two hundred years.

With Islam, this has worse implications than for other cultures that have had rises and falls. For Islam's religious tolerance has always been premised on its own power. It was tolerant when it controlled the territory and called the shots. When it lost territory and saw itself eclipsed by the West in power and civilization, tolerance evaporated. To cite the scholar Bernard Lewis:

> What is truly evil and unacceptable is the domination of infidels over true believers. For true believers to rule misbelievers is proper and natural, since this provides for the maintenance of the holy law and gives the misbelievers both the opportunity and the incentive to embrace the true faith. But for misbelievers to rule over true believers is blasphemous and unnatural, since it leads to the corruption of religion and morality in society and to the flouting or even the abrogation of God's law.

Thus the horror at the establishment of the State of Israel, an infidel country in Muslim lands, a bitter reminder of the eclipse of Islam in the modern world. Thus also the revulsion at former American bases in Saudi Arabia. While colonialism of different degrees is merely political oppression for some cultures, for Islam it was far worse. It was blasphemy that had to be avenged and countered.

I cannot help thinking of this defensiveness when I read stories of the suicide bombers of 9/11 sitting poolside in Florida or racking up a forty-eight-dollar vodka tab in an American restaurant. We tend to think that this assimilation into the West might bring Islamic fundamentalists around somewhat, temper their zeal. But in fact, the opposite is the case. The temptation of American and Western culture—indeed, the very allure of such culture—may well require a repression all the more brutal if it is to be overcome. The transmission of American culture into the heart of what bin Laden calls the Islamic nation requires only two responses—capitulation to unbelief or a radical strike against it. There is little room in the fundamentalist psyche for a moderate accommodation. The very psychological dynamics that lead repressed homosexuals to be viciously homophobic or that entice sexually tempted preachers to inveigh against immorality are the very dynamics that lead vodka-drinking fundamentalists to steer planes into buildings. It is an act of faith. But it is also an act of extreme fear.

V

THIS FEAR DEMANDS AN EXTREME CAPACITY FOR DENIAL. The denial mechanism has to kick in when reality disproves the inerrancy of the faith. The difficulty with a theology that provides a single moment of liberation—the born-again moment—is that it makes subsequent moral failure all the more difficult to own and accept. If you've been saved, how can you go on to sin again and again? How can evil have reemerged after it has been vanquished? And so fundamentalist Christians often refuse to acknowledge their own fallibility or failings in front of their fellows. Or their acknowledgment follows a brutal psychological zigzag of sin, breakdown, recommitment, and then, inevitably, sin again. When faith is provided by an external source, your internal compass may often be wrong. And the wrong demands correction. This is especially true of those given leadership positions in the fundamentalist universe. How can they acknowledge continual errors or even structural flaws if that acknowledgment could lead to others' disbelief?

And so you reach the astonishing spectacle of senior Catholic bishops and cardinals and popes refusing to accept that their fellow priests have engaged in a systematic and criminal abuse of generations of children. By vesting truth in a cadre of selected men, by placing the possibility of salvation on the shoulders of an infallible hierarchy, Catholic fundamentalists found it impossible to concede that some of these very men were moral monsters, rapists, and molesters. The admission of such things would lead to "scandal" and scandal would undermine faith. The higher truth of that faith trumped the lower truth of its practical betrayal. And in that cycle, the authoritative command was to cover up, to insist that what was demonstrably true was untrue, or to threaten and bully and bludgeon deferent parishioners into silence. This is not just a function of a self-protecting clerisy, although that was part of the crisis. It is the product of an entire mind-set that has elevated the human Church to the status of infallible authority and so cannot recognize or correct error within itself. To concede institutional moral failure is to risk the entire edifice of absolute truth that fundamentalists require to sustain themselves. If forced to choose between that truth and the welfare of a single vulnerable child, the hierarchs did what they had to do. They protected the structure of their truth, while violating its actual content.

Similarly, the notion that there were and are many gay priests—perhaps up to a third of all Catholic pastors in the

West—simply could not be acknowledged by the hierarchy. Such an acceptance would again be scandalous and expose the Church to a human inspection of things only God could fully order or judge. And so the existence of gay priests was simply denied; very few were given any counseling or support; the rest were ordered to remain silent about their own orientation. Such a "problem" could not exist, according to the theory, and so it did not exist. Period. And those who were psychologically disturbed enough to be pedophiles or ephebophiles were shunted around in order not to solve the problem, but to hide it. In retrospect, it's astonishing how successful this system was for so long. We have no idea how far back child abuse went in the Church, because so many generations were intimidated by the absolute authority of their clerical superiors into silence or mutual cover-up. What mattered, through long, dark ages of abuse, was the maintenance of appearances, of a doctrine deemed unchangeable and infallible, of a certainty that made empirical reform or openness to conceding error all but impossible.

You can see similar patterns among fundamentalist Protestants. Bawer notices a theme in a book called *Growing Up Fundamentalist* by Stefan Ulstein. Bawer observes:

> One after another, [Ulstein's] interviewees describe families that "stressed the *appearance* of perfection," that were "never interested in getting intimate and honest," that "lived in a world of denial and appearances," that "they don't know how to ask for help" because "they have to appear like they're in charge." One woman speaks of how her mother "can't consider the *possibility* that she made mistakes. It threatens her whole world." Another says: "Everything they do is for show, or, as they would say, a good witness. . . . No one asks for help, because it would blow their cover."

It is too broad a brush to describe all fundamentalist Catholics and Protestants this way, let alone most nonfundamentalist Catholics and Protestants. Many have adopted much more open ways of discussing their human failings, and the most successful have put practical counseling to deal with human problems at the core of their ministry. The success of books like Rick Warren's *The Purpose-Driven Life* is based on his candid and humane acknowledgment that many Christians experience drift, bewilderment, and self-doubt. James Dobson's folksy homilies on how to rear children fall into the same category. Similarly, many contemporary Catholic priests offer compassionate, pastoral care for their flawed flocks and have learned to subject themselves to careful moral scrutiny. Nevertheless, within the logic of fundamentalism you can see the pressure to maintain a facade of perfection for a faith that has to be infallible in every particular. You can see how leaders in such efforts have to project an image of certitude and inerrancy if their certain, inerrant doctrines are to be credible. The brokenness of human beings is much more easily pointed out in the "other" than in oneself and that can lead to the pathology in which you remain in denial about your own flaws, while extremely well attuned to the failures of others.

And so you have the spectacle of Christian fundamentalism being strongest in many regions of the country where abortion, teen pregnancy, and STD transmission are the highest. And you have the fact that the least fundamentalist and most tolerant of states, Massachusetts, has both civil marriage rights for gay couples and the lowest divorce rates in the country. This is not dispositive proof of a connection, but it is suggestive of a dysfunctional relationship with reality. You cannot resolve problems until you acknowledge them. And when you follow a faith that argues that you are already saved, or that the hierarchy can never be wrong, there is a strong temptation to embrace denial. The ultimate Truth outweighs the temporal, human truth of our own weakness and fallibility. How many times have you heard a pope talk about his own errors? Or an Islamist mullah, for that matter?

But that is the role that fundamentalism plays—and has to play. It affirms faith, promises to abolish fear, and offers a clear answer to the bewilderment of modernity. It offers a way to make sense of a world now so fractured in meaning that it appears to defy any other comprehensive explanation or resolution. And this, after all, is what salvation means. It means being saved from one's own reality and released into another one. And that new existence is not a delusion or a dream—but as real as anything in the world. If the facts resist that new reality, then the facts have to be abandoned, not the faith.

Such faith demands adherence and complete obedience. More to the point, it is the act of obedience that must precede any attempt to reason through faith, because reason without faith is insufficient. If your conscience tells you that something the religious authorities or biblical interpreters believe is untrue, it is your obligation to change your mind, to bend your conscience to the will of the church or the Bible. For many Protestant fundamentalists, it requires being "born again," a radical reorientation of a person's life. And here you begin to glimpse the profound difference between the fundamentalist and the conservative temperament. The fundamentalist sees his pre-Christian past as something to be rejected completely. The very phrase "born again" implies the most radical disruption in a human life imaginable—and celebrates it. There can be no nostalgia here, no incremental personal narrative, no attachment to the human past in the face of the divine, because the divine demands totality of commitment. Our stories can be understood only in the context of a larger, greater story called Truth.

For the born-again, there is the life before one is saved and there is the life thereafter. Every fundamentalist life pivots around this bright line. For the conservative, there is merely life—life as a continuing narrative, both personal and communal, a narrative that, like a conversation, has no pre-ordained end and no ultimate truth but the one we give to it. If conservatism is about preserving one's own past, fundamentalism is about erasing it and starting afresh. If conservatism is about the acceptance of imperfection, fundamentalism is about the necessity of perfection now and forever. If conservatism begins with the premise of human error, fundamentalism rests on the fact of divine truth. If conservatism is about the permanence of human nature, fundamentalism looks forward to an apocalypse in which all human nature will be remade by the will of a terrifying and omnipotent God. If conservatism believes in pragmatism and context to determine political choices, fundamentalism relies always on a book. Or rather the Book. And in that political project, as we shall see, fundamentalism has already made great strides.

CHAPTER 3

The Theoconservative Project

"Sides are being chosen, and the future of man hangs in the
balance! The enemies of virtue may be on the march, but they
have not won, and if we put our trust in Christ, they never will. . . .
It is for us then to do as our heroes have always done and put our
faith in the perfect redeeming love of Jesus Christ."

—Tom DeLay,
former Republican House Majority Leader, March 2006

I

IN TOTALITARIAN OR AUTHORITARIAN SOCIETIES, THE
fundamentalist has an obvious option to realize his dreams.
He merely has to control the government, by force if necessary. That's what the Ayatollah Khomeini did in Iran in 1979;
it's what the Taliban did in Afghanistan at the end of the last
century; it's what Oliver Cromwell did in seventeenth-century
England. In modern constitutional societies like America, however, the difficulty is much, much greater. The First Amendment guarantees freedom of religion, which also implies
freedom from religion. The Constitution protects religious and
other minorities from untrammeled religiously inspired majorities. And so Christian fundamentalists in public arguments in
contemporary America find themselves forced to invoke merely
their personal relationship with Jesus Christ to persuade others
of the rightness of religious solutions to political problems. But
what if their interlocutors don't share that personal relationship?

How can the fundamentalist psyche make real political inroads through democratic debate with nonbelievers or nonfundamentalist Christians or Jews?

In one respect, of course, they don't have to. Merely by aligning fundamentalist churches and populations with a single political party, fundamentalist Christians have been able to wield enormous influence without having to persuade anyone else at all. By selecting Supreme Court justices less willing to step in and protect minorities from majority power, they can also advance their agenda. But they still can't muster clear majorities; and relying solely on religious authority—whether the pope's or the Bible's—puts them at a great disadvantage in the critical battle of ideas. What they need is a more rigorous public philosophy, one that can provide the intellectual muscle behind populist religious appeals. Merely invoking Jesus or the Bible or the pope in multicultural societies doesn't cut it anymore.

Where to go? As they grappled with increased secularism in the 1970s, the fundamentalists realized they required a real philosophy, a real argument based on reason to advance their cause. Where to turn? There were and are options. If the first mistake with respect to fundamentalism is to underestimate its appeal, the second is to undervalue its intellectual coherence. The power of fundamentalism is precisely this coherence. It makes sense of everything—and marinates that sense in the power of God's love. Nonbelievers who think that fundamentalism is merely an existential leap or an emotional crutch often forget that fundamentalisms of various stripes, from Islam to Christianity, have developed over the centuries rich, careful, nuanced accounts of human nature, conduct, and morality. As their doctrines have come up against logical gaps, they have been filled in and finessed.

In their most highly developed and evolved forms, these theories are not the equivalent of thumping a Bible as if its very existence answered critical human questions. They are not mere invocations of the authority of a text, a pastor, or a personal relationship with Jesus. They are, in a word, *arguments* about human life and meaning that necessarily have to give an account of the reality we all experience; and so these arguments are subject to analysis, argument, debate by nonbelievers as well.

There is a reason, after all, that the great religions that generate the fundamentalist mind-set today predated and outlived the secular fundamentalisms of the recent past. The totalitarianisms of the last century—the economic determinism of communism and the racial eugenicism of Nazism—were forms of fundamentalism. They too answered every question and penetrated every corner of the lives of those who lived by them. But they perished in the relative blink of an historical eye—their manifest untruths exposed with, by historical standards, remarkable rapidity. The great faiths of Judaism, Christianity, and Islam, in contrast, have already lasted centuries and show few signs of collapse. And one reason for their durability is that they are better thought out, they encompass profound truths about the human condition, and, most tellingly, they have proven themselves by the test of the millennia.

Contemporaries who dismiss ancient truths because they are indeed ancient merely display their own intellectual laziness. Bad ideas do indeed perish over time, as human experience refutes or undermines them. Crude ideas, in turn, are often refined by the ages, as arguments and debate and occasional flashes of individual genius, change our view of things, and shape a philosophy the way tides shape pebbles. *Of course* these religious traditions have outlasted the crackpot theories of a single man, Karl Marx, or the hodgepodge of racist theories and anti-Semitic frenzy that defined Nazism. They have undergone rigorous testing over centuries, and, despite many changes and evolutions, something in their core message has endured.

One such tradition of religious argument is known as "natural law." Struggling to find a way to win the battle of ideas, Christian fundamentalists have turned in frustration in the last few decades to this ancient but robust tradition of moral thinking as a way to persuade their opponents in such vital moral debates as abortion, gay rights, end-of-life decisions, contraception, and the family. And when you examine and ponder the arguments of natural law, you can see why it still has such appeal—especially to fundamentalists in search of reasons as well as revelations. Its claims are a sweeping and profound challenge to the fragile pluralism and secularism of the current West. If it is true that conservatism has shifted in these past few years toward a more explicitly theological basis, then this adoption of natural law arguments is central to that transition. Critics have dubbed that movement "theoconservatism." It is central to understanding the enormous shift in American public policy in the Bush years.

One of the key contemporary proponents of natural law is Robert P. George, of Princeton University, a Catholic political theorist, who, along with other public intellectuals such as Richard John Neuhaus and George Weigel, played a key part in creating the theoconservative movement and integrating it into mainstream Republicanism. These are not fringe figures, but central ones, especially in any attempt to understand the revolution in American conservatism that has taken place in the last few years. Neuhaus—or "Father Richard," as president

Bush routinely calls him—is one of the few public figures with access to both Karl Rove and Pope Benedict XVI. George has constructed a career around grooming theoconservative scholars and political activists from a well-financed base at Princeton University. All the key conservative journals of opinion in America—from the *National Review* to the *Weekly Standard* to *First Things*—now have theoconservatives installed as arbiters of the conservative debate on social policy.

Neuhaus is a former leftist who came to believe in the 1970s that American constitutional democracy could not survive without being reformulated around a religious core. He longed at one point for a Christian Marx to revolutionize America. As the writer Damon Linker has argued, Neuhaus wanted an ideological "alternative both to Marxism and secularized liberalism" that would give America "a definition of reality, an ideology, based on Jewish-Christian religion, that [was] as creative, comprehensive, and compelling as was Marx's definition of reality." In this project, reason was to submit to the authority of the pope, who would be required to lead such a newly invigorated Christianity. "I think for myself not to come up with my own teaching," Neuhaus explained, "but to make the Church's teaching my own." The role of reason was to provide public arguments for the truth of the church teaching; but ultimately, that reason had to obey divine authority.

Weigel has gone even further, asserting that the radical decision of America's founding fathers to separate church and state was in fact based on a Catholic understanding of the relationship of politics to eternity. But the central question for Weigel is less whether Catholicism is "compatible with democracy" than whether America, or any Western form of government, "could survive unless it reconstructed a public consensus around those 'elementary affirmations' upon which it was founded—affirmations whose roots . . . were not the original product of the Enlightenment and its American deist heirs, but of the Catholic medieval theory of man and society."

And so on the bones of an evangelical revival, and an alignment of one form of religion with one political party, the theoconservatives provided the meat of philosophical argument. They brought a brain to the fundamentalist psyche; and made arguments about social policy that, in the 1990s and new century, helped tear the United States apart at the seams. If you want to know why abortion has come to dominate American political debate, why homosexuality has been thrust into the center of the culture wars, why the Terri Schiavo case became such a pivotal moment in American history, and why sex itself is now inextricable from political warfare, then it is important to understand the arguments these men have made.

II

WHAT IS THEIR FUNDAMENTAL ARGUMENT? I WILL DO what I can to paraphrase it briefly without doing too much damage to its nuances. It is that human beings have a common, fundamental nature by which they should and must be judged. That nature is given by the Creator and is the arbiter of how we are supposed to live our lives. Because it is our nature, and easily observable to even the untrained but open mind, it can be shown to exist by argument and proof. Science will support it, because science cannot conflict with truth. Moreover, this nature points to a purpose for human beings through and past their passing wants and needs. From the "is" comes the "ought"—or, rather, the distinction between an "is" and an "ought" is specifically challenged.

The modern idea that we have a given biological, evolutionary, physical DNA, but that this DNA gives us no real insight into how we should live our lives, or what our morality ought to be, is an idea the theoconservatives explicitly reject. After all, they argue, if God made us, He made us for a reason; and our goal on Earth must be to understand that reason and to do our best to achieve it. Although each of us may have a different specific role in life, different abilities or talents, we all share basic human nature and so share basic human goods. The good life is the same for all; the virtues apply to all and they are imprinted on our nature, asking merely to be fulfilled. So an observant Catholic who inerrantly follows the "laws of nature" experiences no conflict between, say, her desires and her morality. They both ultimately come together because God cannot make creatures that are internally incoherent, that are biologically wired to render the achievement of virtue impossible. Everything has a purpose, human beings most of all. All we have to do is to examine our own, shared human nature to see what that purpose is.

An important caveat: What is natural isn't, from this viewpoint, simply what is found in nature. If it were, it would be disproved by every exception to the rule. A dog that is born with three legs as a genetic defect doesn't redefine "dogness," the nature of dogs, the purpose of dogs, or what it is good for a dog to be at its best. Its exception proves the larger natural rule. A bird with damaged wings does not redefine the nature of a "birdness" as something that does not fly; and yet it is still a bird. Similarly human beings born with mental disabilities or dying in the haze of late Alzheimer's disease may be in accordance with nature as it actually exists, but not in accordance with these individuals' own fully realized nature as human beings. And fully realizing your nature is a form of flourishing, and in the case of rational, self-conscious beings like humans, also an inextricably moral form of flourishing. There is a good and a bad way of being human, the natural law theorists argue. And the good way is more fully human as well as more moral than the bad. It brings out our nature in its most developed and moral form, fosters happiness on earth and makes possible salvation in heaven.

Here's a pretty clear statement of the argument, in Robert George's own words:

> The Church teaches that a person's nature, in the sense relevant to moral judgment, is constituted by human goods that give him *reasons* to act and to refrain from acting, and not by *desires* that may, rightly or wrongly, also provide motivation. These "natural goods" are "basic" inasmuch as they are ends or purposes that have their intelligibility not merely as means to other ends, but as intrinsic aspects of human well-being and fulfillment.

Notice a few things. First, our nature has a rational core. It is not some random accumulation of DNA, inherited by millennia of natural selection. It is a nature imprinted with reasons known by God and accessible to humankind, and geared toward a purpose. Second, a human being who follows this purpose and fulfills his own nature, experiences a personal coherence and well-being that releases him from internal conflict or persistent sin and allows him to be more fully human, to realize his fullest potential as a member of the species *Homo sapiens*. Third, an individual's own account of her own experience, what she feels or senses or knows about herself, is ultimately irrelevant to her own correct self-understanding. Just because an individual human being may find herself drawn irresistibly toward men who have commitment-phobia does not mean that her nature is best realized by marrying one of them. Reason contradicts desire; and the moral person tries hard to conform her desires, even the most fundamental ones, to the form reason dictates.

The critical feature of all fundamentalisms is present in this theory: this is a total system in which everything is explained and everything is capable of ultimate harmony. Our only choice is whether to live against the grain of this nature—or in accordance with its patterns and direction. In this choice lies not only the difference between being saved and being damned for eternity; but also the difference between being happy and being miserable on planet Earth. Again: There is no conflict here, no notion that being moral on Earth will make one unhappy for the present but ripe for salvation. Everything is congruent, and everything makes sense, except of course that nature sometimes throws obstacles in the way of the fully realized human life. Poverty, disease, accident, genetic variation: all these can impede but never eclipse the possibilities of being a fully flourishing human being—in a moral, spiritual, physical, and intellectual sense.

This argument has its roots in Aristotle's account of human nature and virtue, with Aquinas's Christian theology tacked on and integrated into the analysis. It may seem abstract when put in general terms, so let's get into the nitty-gritty. Let's take up a central question in our debate about values, something that has provoked court cases, roiled political campaigns, and dominated cable talk shows: sex.

I don't raise this issue right away for prurient or personal reasons. Any theory of human nature will have to deal with the mechanism by which humans reproduce themselves. Such a basic function is critical to the existence of the species at all—up there with breathing and eating—and requires understanding as much as any other facet of our lives. It is also a central front in what we have come to call our culture wars, and so helps illuminate how these philosophical ideas have come to affect national politics. It therefore offers a fascinating account of exactly what is at stake here. It helps us see how exactly natural law affects the way we live and the judgments we make about our actions.

The conservative professor Edward Feser offered a very clear statement of natural law's basic case about sex on his blog in May 2005. I reproduce his argument here because it's as straightforward a version as any I have read, and a little less encumbered by abstractions than George's. Feser looks at sexuality and asks, as any natural lawyer would, what it is *for*:

> If you consider the sexual drives that human beings have, then it is blindingly obvious that if those drives have any natural purpose at all—if they were, say, designed with a certain end in view—then that purpose is to get people to use their sexual organs. And if you consider the sexual organs themselves, then it is also blindingly obvious that if they were designed with any purpose in mind, then that purpose is procreation. More specifically, the purpose of a penis—again, if you assume that it was indeed designed with a purpose in mind—is quite obviously to deposit semen into a vagina (and also, of course, to urinate). That's what it's for, if indeed it is for anything, and whether or not it can be used for other purposes. You can use a corkscrew for all sorts of things—cleaning your fingernails, say—and you might for some reason even have a compulsion to use it only to clean your fingernails. The fact remains that what a corkscrew is for is opening bottles. And the purpose of sexual organs, if they have one, isn't any more mysterious than that of corkscrews.

It is not easy to dismiss this argument out of hand. Human beings intuitively infer what something is for from what it can do. And *if* you believe that human nature has a purpose, then it would certainly be odd to believe that a penis and a vagina had nothing to do with reproduction. At the same time, it's obvious that many aspects of our bodies have varied purposes or no discernible purpose at all. In the very example Feser gives, the penis is designed both to urinate and to ejaculate. How does one judge which of these functions is the core one? Urination is far more common than ejaculation. While a human male can be sterilized and render the penis incapable of producing semen, he cannot easily relinquish the urination function if he is to survive. In that sense, the penis is really for expelling toxins, not making babies. But this merely makes us reflect on the occasional silliness of insisting on a singular purpose or role for any particular organ. The body is far more complex and interconnected than reducing any one of its parts to a single purpose. Our tongues, to take another random example, help us speak and they help us detect what is good to eat. But they can also be used to insult someone, to lick one's lips, or to give sexual pleasure. Which of these things are they for? What is the fundamental nature of a tongue? What is it for?

Feser tries to help us with these questions by offering a further nuance:

> It must also be emphasized that, contrary to another common misunderstanding, "unnatural" in the context of the view I'm describing does not mean "using something *other than* for its natural purpose." It means "using it in a manner *contrary to* its natural purpose." . . . There is nothing unnatural about merely tapping out a little song on your teeth, even if that's not what teeth are for. But there is something unnatural about

painting little pictures on your teeth and then refusing ever to eat again lest the pictures be rubbed off, or pulling them out so as to make a necklace out of them. The former sort of act does not frustrate the natural end of teeth, but the latter acts do. And part of the idea in the traditional natural law understanding of the sexual act is that ejaculating into a Kleenex, or a condom, or into any bodily orifice other than a vagina, doesn't just involve using an organ other than for its natural purpose (which is not necessarily "unnatural") but that it uses it in a manner contrary to its natural purpose. For the "aim" or point of arousal and ejaculation, if they have an aim or point at all, is to get semen into a vagina.

Let's concede the premise of Feser's and George's argument and see where it leads us. Let's assume that our bodies, as natural lawyers argue, are inextricable from our souls and minds, form a seamless natural, coherent whole, and are indeed designed for a purpose. Let's also assume that it is possible to use the body in ways that are "contrary" to that purpose. Does Feser's argument about sex hold up?

Let's posit a perfect, Catholic married couple who live their life according to natural law in every respect. They never use contraception and never engage in any sexual act that does not result in the penis depositing semen in a vagina. Assume further that both are fertile. If the woman conceives, then it must follow that the point of depositing semen in the vagina is literally irrelevant for the following nine months or so. Exclusively procreative sex, in other words, *naturally* necessitates continual nonprocreative sex. If the husband were to refrain from any sexual activity in those nine months, to avoid activity "contrary" to nature, he wouldn't stop producing sperm. Even if he did not masturbate, his body would emit sperm in nocturnal emissions, triggered by erotic desire programmed into his brain. In other words, his body itself would spontaneously and without any intention on his part act in a way that is "contrary" to the natural law. It's hard to see how such a phenomenon can be deemed "contrary" to any reasonable account of "nature."

In fact, any reasonable account of human sexuality, observed by any rational person, would conclude that, even under the strictest of Catholic moral teachings, most sexual activity between a man and wife must necessarily be nonprocreative. A man's sex drive does not disappear the moment his wife becomes pregnant, and neither does his wife's. Put it another way: If the purpose of the penis was solely procreation with one monogamous female partner, as the natural law philosophers insist, it would surely have been designed very differently. It would ejaculate much more rarely and be attuned to another human's menstrual and reproductive cycle. To take this thought experiment still further, it might even be designed only to ejaculate by some mechanism within the vagina and by no other means. In fact, of course, the male sexual organs produce an almost infinite number of sperm, an infinitesimal amount of which will ever become another human life. It seems odd that when a phenomenon occurs to one sperm out of millions, the one should be seen as the rule and the millions the exception. Moreover, the penis can be stimulated by almost any other physical object, and at its sexual peak, may ejaculate countless sperm several times a day. Why would this be a fact of our physical nature if the sole purpose of a penis is to procreate with one other human being? These are not arguments. They are facts. If nature implies a purpose, then the purpose of the male sexual organs posited by natural law is obviously divorced from its actual, empirically observed function.

The same, of course, applies to the female body. Pregnancy removes the female from reproductive sex for long periods of time. The primary source of sexual pleasure, the clitoris, is also biologically removed from the act of procreation. A man can successfully impregnate a woman without ever stimulating her clitoris or giving her sexual pleasure. A man can successfully impregnate a woman by raping her. The very existence of the clitoris is therefore a living rebuke to those who argue that nature itself—the way our bodies have been constructed—dictates a certain and necessarily procreative sexual morality. To echo Feser's argument, what is the proper use of the clitoris? It plays no essential role in actual procreation, and yet is the prime source of female sexual pleasure. If this isn't an indication that nature allows for sex purely as pleasure or as a pleasurable way to intensify and deepen an emotional bond, then what is? Or to take a second example. All women eventually stop menstruating. On average, they live longer lives than men, and so whole swathes of their lives entail, as a function of biology, that their sex lives will be consciously and unavoidably nonreproductive. If natural law is premised on a commonsensical inference from nature, then how on earth can it be construed not merely to argue that procreation is the sine qua non of sex, but that everything else is anathema?

One answer to this question comes from the revamped natural law of recent decades. In this recalibrated argument, it is also "natural" that one man and one woman develop and deepen a "unitive" emotional bond. This is good for them and good for children and good for society. It also resonates with a deep longing in both men and women—albeit to very different degrees—for emotional bonding, security, and stability. Since all sexual acts during pregnancy and after menopause by definition cannot be procreative, this argument plugs a gaping lacuna in the case. These nonprocreative acts can be described as natural because they help deepen and foster the emotional bond that is the natural end of a fully flourishing human being. But if sex can be merely unitive in its natural function, then why not sex with contraception, mutual masturbation, or oral and anal sex? If nonprocreative but unitive sex in a loving relationship is intended to deepen the marital bond, keep a family within manageable economic limits, and help buttress, rather than undermine, family life, why is it "contrary" to natural law in all cases? This is the very reasonable argument most Protestants make, which is why they have come to accept contraception and even divorce, despite unqualified biblical condemnation of the latter. It is also why large majorities of Catholics also regard the Vatican's insistence on this matter unconvincing.

George and his allies have an answer to this. They argue that intentionally nonprocreative sex is moral in marriage as long as it is unavoidable and still *mimics* the procreative act. They accept—and this is a relatively new development—that sexual activity by nature is designed to deepen and foster unitive love between a male and a female, and that the enrichment of this emotional bond is also integral to the sexual act. So when actual reproduction is impossible—when a wife is pregnant or postmenopausal—this "unitive" function does the work of the "procreative" function integral to a moral sex life. What matters, even if procreation is impossible, is that the symbolic act of procreation be practiced.

This is an intriguing concession, but one that opens a lot of further questions. If it's possible for nonprocreative sex to be natural even when it is not designed to reproduce and may not even be able to reproduce in fact, then why can this not be extended as a general rule? If a woman has had, say, four children, and sincerely believes that more would lead to great financial stress, emotional exhaustion, and strain on her unitive bond with her husband, why couldn't she use contraception and have sex as a purely unitive rather than procreative act? This is the sensible compromise reached by many Protestants. And it is the de facto compromise of the vast majority of Catholics. Why is it impermissible for the natural law fundamentalists?

The answer reveals that we have entered a circular world of theological logic, rather than an open-minded deduction from observing the natural world. The answer is that humans are designed in such a way that all their sexual expression must either achieve or, at the very least, "mimic" the authentic natural act of reproduction. But why mimic, even when there is no chance for actual consummation? The answer, it seems, is that the perfect is the standard by which everything else must be judged. In this respect, the perfect is not merely better than the good, but actively intolerant of it. There is one model for human sexuality—reproductive heterosexual intercourse—and it is a jealous model. Even though the vast majority of sexual acts are nonprocreative, the symbolism of reproduction must always be present. That symbolism is bound up in the notion of humanity being gendered into male and female as a core feature of nature itself. And that symbolism, instantiated in the civil or religious ritual of marriage, can never be compromised.

For the followers of natural law, there is no distinction between a human being's mind and his body, his will and his desires. Both are part of a completely indivisible whole, imprinted with a nature and design by God. If a human being has sex consciously designed to avoid reproduction or experiences sexual pleasure or function unrelated to reproduction, his will is violating his nature. He is not being one with his body; he is using it as an instrument for his own pleasure and desires, even if he is doing so as a means to deepen a committed emotional bond. He is literally dividing himself into two parts: the natural and the unnatural. And so he destroys his own human integrity.

Here's Robert George on this point, quoting from another natural law scholar, John Finnis:

> Marriage, according to Finnis, is one of the basic human goods. As such, it provides a noninstrumental reason for spouses to unite bodily in acts of genital intercourse. This bodily union is the biological matrix of the multi-level (bodily, emotional, dispositional, and even spritual) relationship that is in marriage . . . [Thus, marital] acts realize the intrinsic good of marriage itself as a two-in-one flesh communion of persons . . . If this view, or something like it, is sound, then it is plain that oral or anal sexual intercourse, whether engaged in by partners of the same sex or opposite sexes, and, indeed, even if engaged in by marriage partners, cannot be marital . . . Moreover, masturbatory and sodomitical acts, by their nature, instrumentalize the bodies of those choosing to engage in them in a way that cannot but damage their integrity as persons.

George then quotes Finnis, describing what happens when a married couple has oral sex: "[T]he partners treat their bodies as instruments to be used in service of their consciously experiencing selves; their choice to engage in such conduct thus dis-integrates each of them precisely as acting persons." It's important to see here that it matters not at all what the conscientious intentions of the partners are. Even if they believe such acts are actually helping cement their relationship, they are objectively destroying it. Even though they believe that they are able to make moral choices in this area by consulting their own consciences, those consciences are subject to objective moral law that makes such choices illicit. Even if they believe that their decision helps them retain the integrity of their marriage and the profound experience of sexual pleasure with a loved one, they are in denial about the damage they are doing to their own bodies and souls. In short, these natural law philosophers deny any distinction between mind and body, they deny any preeminent role for conscience, they merely assert that nature demands unitive procreation as the sole purpose of sexuality, and they demand no deviation whatever from the rule.

When a reasonable person cites natural evidence against this argument—the menstrual cycle, the overwhelmingly superfluous production of sperm, the placement and role of the clitoris, the nonprocreative periods of pregnancy—she is referred by the natural law philosophers to a deeper "nature" that is somehow truer or more natural than the nature we can all observe or understand. And here is surely the weakest link in the natural law case. The basic assertion of natural law philosophers is that their argument is self-evidently true to any reasonable person. But quite obviously, the vast majority of human beings who have ever lived, and even the vast majority of Catholics, seem to have a different idea. Are these people being completely irrational?

I have just given several reasonable arguments that, at the very least, suggest a more complex view of human sexuality than the narrow and constrained doctrine of natural law. These counterarguments, moreover, accept the premise of natural law: they are based on direct and reasoned observation of what is in nature. It seems to me that reasonable people can indeed differ on the nuances of what is natural or unnatural, moral or immoral, in human sexual expression. And as long as such reasonable differences exist, then the power of the natural law argument collapses. If one view is not self-evident, then other views can be deemed as legitimate inferences from nature. To cling to the purism of natural law's insistence on exclusively procreative or pseudo-procreative intercourse requires making a secondary argument that cannot be inferred merely from observing nature. It requires an argument about God's will, or the moral danger of sexual pleasure in itself, or the inability of complex intelligent beings to have a range of sexual expression, rather than a single, exclusively moral one.

But this insistence on a unitary, exclusive form of sexual expression is what one would expect from the fundamentalist mind. The fundamentalist does not tolerate a diversity of views. There is one truth; and all other pretenders are threats to it, or contradict it. The notion that our sexual natures may have multiple purposes, or legitimately encompass manifold forms of expression, or that there may be a hierarchy of moral sexual acts, with some nonprocreative acts being a moral background to the highest form of sexual expression: these complex compromises are precisely what fundamentalists refuse to live with. If you concede the principle that sex can be morally divorced from procreation, then fundamentalists lose the rock on which their entire edifice is built. And so they do not merely differ from others on the role of sexuality, they aggressively oppose all alternatives. They argue not only that one form of sexual activity is preferable to others, but that all others are immoral, disintegrate the human person, and violate God's law. The deeper you examine this argument, the more rigid and contrived it becomes, the more divorced from nature it appears, and the more aggressively controlling of human freedom it seems.

This is not to say it doesn't offer a valuable insight. The insight that sex is indeed, at its root, linked to biological reproduction is undeniable. The complete severance of the two is worth debating and pondering. It is also true that sex is connected in a profound way to human longing for a union, spiritual and psychological, with another person. These arguments are valuable, profound, and even beautiful in their description of the human experience. And they could form the basis of a rich and rewarding account of human life and sexual morality. But the Catholic fundamentalists, like other fundamentalists, are simply not content to leave the matter there, or to explore its possibilities further. They have to condemn all other variations of sexuality—in practice, the overwhelming majority of sexual acts even within faithful marriages—as counter to human nature, sinful, and even subject to legal sanction. In this endeavor, we can only say that they are unconvincing on their own terms. Even if you posit that sexuality has a purpose, even if you concede that nature and biology are the central guides to that purpose, even if you believe that sex is designed for a monogamous couple dedicated to a shared life together, it is still illogical to restrict moral sexual behavior exclusively to the extremely narrow confines that George, Feser, Finnis, and others insist upon.

III

SO FAR, OF COURSE, I HAVE ADDRESSED THE NATURAL LAW arguments on their own terms. But it would be derelict not to note that almost no contemporary biological scientists even come close to viewing human nature in this way. A philosophy derived, with some haphazard modern patches, from the natural history of an ancient Greek and an early medieval scholar can be forgiven for ignoring the scientific revolutions that have occurred in the past few centuries. But the overwhelming scientific view is that sexual reproduction has to be seen within the prism of Darwinian natural selection; that its goal is to diversify the human genetic pool to render it more resistant to changes in its environment; and that an integral part of this program for genetic survival is a massive superfluity of sexual life and activity.

Men produce exponentially more sperm than will ever become future humans. Why? The only convincing reason yet proffered is Darwin's: Because such an excess of production maximizes a human male's chances of successful reproduction. Integral to this dynamic is the assumption that almost all of this sperm will go to waste. The same applies to the superfluity of female eggs. The point is precisely a sexual nature whose purpose may ultimately be reproduction but which has built into its own nature a huge reserve of wasted sperm, discarded eggs, miscarried embryos, and a constantly churning lottery of genetic splicing and mixing that serves to maximize the chances for human survival and the passing on of diverse DNA. It is a very, very messy system that treats individual human life with casual indifference as it advances its overall goal of propagating the species as a whole. In other words, nature itself as we have come to understand it, points far more emphatically toward the ubiquity of nonprocreative sex as the objective norm of human sexuality.

This strategy for species survival implies something else as well. It suggests that males are genetically programmed to have sex with multiple partners, not just one, and that females are genetically programmed to find mates who are more likely to provide healthy offspring and to hang around after the babies are born. The founders of natural law—Aristotle and Aquinas—were as unaware of these dynamics as they were ignorant about female anatomy. But any fair reader of those great philosophers would be in no doubt that they would have been fascinated by the uncovering of new worlds and new theories. These philosophers were trying to make rational sense of the natural world around them, and, in Aquinas's case, to help it cohere within the context of Christianity. The idea that these philosophers would not want to know what we have subsequently found out about human nature misreads their seriousness and open-mindedness. It is to turn their arguments and observations into others' doctrines and assertions, to freeze human thought at a particular moment in its development, and to blind oneself to new and revealing facts. Indeed, you can read any number of essays by natural law theorists and find not a single reference to the scientific literature of the past five hundred years, let alone the past fifty. And yet they rest their case on what is found in nature. This is ideology, not reason.

Now it is equally true that scientists, at least the reputable ones, derive no moral inferences from Darwinian theories. That's not their business. Indeed their business is premised on an understanding of science that has no interest in good or bad but merely correct or incorrect. A Darwinian may well believe that natural selection shows how we come to be what we naturally are, but he may also strongly believe that our morality may require us to resist that nature or even overcome it. So the naturally promiscuous male may be required by morality to violate his basic nature and dedicate himself to one woman to the exclusion of all others. In some ways, of course, that argument is also derived from nature—but it suggests that nature is a force to be understood and resisted, rather than imagined and followed.

This is not the right occasion to unearth the entire debate about natural selection. But very few reasonable and educated people today can assert that "nature" somehow dictates certain moral behavior, without doing enormous damage to the meaning of nature and the meaning of moral. And that, in fact, is what the natural law theorists are doing: they are smuggling in a fundamentalist doctrine under the guise of empirical analysis. They are welcome to their doctrine as a function of revelation or philosophy. But as the only reasonable way to infer morality from nature? On its face, it collapses as an argument after the briefest of inspections.

But it's also a revealing argument about the deeper mind-set of fundamentalism. Fundamentalists assert a central core idea and then contort or distort reality in order to make it fit their model. They take the ideal and insist that it be universal. In sexual matters, they rightly describe the extraordinary, irreplaceable moment when a married man and woman consummate their love by creating new life—but then refuse to allow that sexuality as a complex human reality can be morally experienced in other ways as well. They see exceptions to their rule not as invitations to explore more about the complexity of human and natural life, but as a way to condemn and exclude any activity that doesn't fit that ideal.

Yes, a three-legged dog may not represent the ideal form of dogness. But that is no reason to forbid it from running to catch a ball. Yes, a loving couple may experience their deepest and most meaningful sexual moment when it includes the conception of another human being. But that doesn't mean that the nonprocreative part of their sex life doesn't meaningfully address physical, emotional, and spiritual goals that are worthy and pleasurable in and of themselves. The natural law philosophers seem to believe that human pleasure can only be licit if harnessed to a greater purpose, rather than sought out as an aspect of human experience in itself. They are entitled to these moral views and to the assertions that accompany them. What they are not entitled to is the endorsement of empirical human "nature" as the imprimatur for their somewhat eccentric opinions.

Moreover in their fundamental goal—to find a single, inerrant path to human flourishing and morality—they ignore a critical element of human nature as we now understand it. If the theoconservatives abhor what is "unnatural," the natural world does the opposite. In fact, it moves forward through "error" or variation. It is diversity in genetic replication that creates the openings for evolution and natural selection; exceptions are what help create the natural rule. Genetic difference might give an animal an unexpected advantage in a changing environment, and so, over millennia, become more and more common until it helps define the very nature of the species as a whole. This is how evolution works—not through a perfect design, but through a bewilderingly messy game of chance, time, and almost infinite genetic variety.

Changes in environment also mean that some previous genetic adaptations become unnecessary. But they may be buried deep into our genetic code and so stay on regardless. They are natural but pointless—an evidentiary rebuke to those who want to fit nature into a single coherent model with a single, defining purpose. We now know that vast amounts of our human DNA are effectively useless junk—the consequence of random mutation and natural selection through the ages, accumulated for different reasons at different times and much of it sitting there pointlessly—useful for some reason in the fathomless genetic past but irrelevant today.

We know that in nature itself, the rule is immense, bewildering diversity in life's expressions. Our closest genetic cousins among the apes can be observed to have "natural" inclinations in sexual life ranging from relative monogamy to rampant promiscuity. The bonobo chimpanzees, one of humankind's closest genetic cousins, deploy masturbation as a form of casual greeting. By empirical observation, *Homo sapiens* is a moderately adulterous species, made up primarily of mildly unfaithful male-female couples with a small minority of same-sex coupling. This is our nature, imprinted indelibly on our genes. But of course, there is also enormous diversity within the species as well, with some of us having an easier time with monogamy than others; and women seemingly more adept at commitment in general than men. All we can know for sure is that this nature is what our environment slowly over countless millennia encouraged and fostered as the best means for our survival. It may offer some insights with which to infer a sexual morality, but it certainly doesn't dictate it. And even more so, it does not exclude a diversity of moral sexual experience and identity.

Science tries to explain and understand such diversity. Fundamentalism is threatened by it, creates a single model for human flourishing, and then attempts to extirpate or regulate or merely condemn anyone and any activity that falls outside that model. The only way to do this is to ignore basic facts of human biology, discard centuries of scientific research, and stigmatize the very errors and diversity that made human survival possible in the first place.

Take the very real phenomenon of intersex human beings, babies born, for a variety of reasons, with indeterminate genitalia. Given the nature of natural selection, the randomness of genetic splicing, and the complex hormonal factors that can influence pregnancy, such a gray area in gender and sexuality is completely predictable. Does that mean that these people are somehow less people? Or that they are incapable of a moral sexuality? If there can be no distinction made between mind and body, between reason and desire, since we are all coherent, integrated humans, how is it that for some transgendered individuals, their minds and souls tell them they are one gender and their genitalia tell them they are another? It requires a philosophical leap to describe this as a biological "error" rather than as another manifestation of an infinitely complex creation. And that leap is by no means self-evident.

The same absolutism applies, of course, to the phenomenon of homosexuality, the attraction to members of the same gender that a small minority of human beings experience, and, so far as we can tell, have always experienced. For the natural law philosophers, there is no more unnatural a condition. Because they have defined the world in absolutist terms of male, female, and sex-as-reproduction, the very existence of homosexuality must be defined as, in Pope Benedict's terms, an "objective disorder." This disorder is so profound a psychological and moral problem it does not merely bar homosexual persons from any moral sexual intimacy, but also bars even celibate homosexuals from the priesthood. But anyone who has absorbed the overwhelming scientific evidence of the last century and a half will almost certainly conclude that the persistence of homosexual behavior and relationships in every culture and every community in human history suggests that it is indeed part of nature's complicated design.

Some evolutionary biologists have posited a variety of explanations for homosexuality's resilience: communities with homosexuals might have had an evolutionary advantage in retaining a cadre of men without children for the purposes of, say, professional war-fighting, or teaching, or even a priestly caste. These are currently mere speculations. But they come from a mind-set that sees natural differences and seeks to explain and understand them. Fundamentalists, in contrast, begin with an ideal version of human experience and seek to marginalize or stigmatize anything else. And they do so not merely by identifying the exception as rare—as in, say, left-handed people or redheads—but as morally repugnant and alien to human "nature." In this they tell us less about human nature than about the world they want to live in—a world which nature, in an exquisite irony, doggedly denies them.

IV

IN THE EARLY CHRISTIAN ERA, ABORTION WAS REGARDED as immoral primarily because it was a way in which sex could be divorced from procreation. It was a cover-up for adultery or sodomy, seen as a way in which the consequences of illicit fornication could be hidden, and therefore a retroactive form of contracepted sex. This is how Augustine viewed it; and how Aquinas understood its immorality. Abortion, mind you, was still morally wrong in all cases. But it was wrong primarily because it undermined the natural basis of sex as exclusively a means to produce babies, not because it was always deemed the taking of human life. Later medieval Christian scholars took a similar view. In a penitential ascribed to the English monk Bede, the penance for abortion was as follows: "A mother who kills her child before the fortieth day shall do penance for one year. If it is after the child has become alive, [she shall do penance] as a murderess. But it makes a great difference whether a poor woman does it on account of the diffi-

culty of supporting [the child] or a harlot for the sake of conceal-
ing her wickedness."

From the beginning, in other words, the Church took a
somewhat nuanced view of the issue. It judged the morality of
abortion according to the time after conception in which it took
place, the context and intention behind it, and left a huge moral
gray area between the moment when a human person could be
viewed as existing within a mother's womb and when it was
eventually born. There was a critical distinction drawn here between conception and "ensoulment," the moment when a soul
"enters" the body, when human life becomes a human person.
Bede suggested that the timing of that moment was forty days
after conception. Aquinas argued that the soul was bestowed
on the fetus when it had developed into a body proportionate to the need for a rational guiding function, that is, a mind
or soul (although Aquinas thought that this function rested in
the heart). Like Bede, Aquinas put this moment at forty days
for male fetuses and more than eighty days for females. That
strange gender discrepancy alone speaks volumes about the state
of biological knowledge in Aquinas's time. They were making it
up as best they could.

This was the consensus for the vast majority of Christian
history. Abortion was never regarded as defensible, but the
gravity of the sin was dependent on a variety of extenuating
factors: the precise development of the fetus, the motives of the
mother, and the circumstances in which the pregnant woman
found herself. In 1140 one of the earliest formal attempts to
codify Church law on the subject ruled that abortion was not
homicide if it occurred before the formation of an actual fetus,
when the unborn life was what we today would call a zygote,
blastocyst, or embryo. Aquinas reasoned that unborn life went

through three stages—a vegetative period, an animal stage, and finally a rational moment when the basis for cognition and reason could be detected, and "ensoulment" fully recognized. It's still striking that even for Aquinas, this process happened very quickly. There was only a little less than a month and a half after conception for human life to exist in a separate category from a human person.

Nevertheless, when you compare the concern, pragmatism, and empirical care of early Christianity with today's absolutist insistence that all abortion at any moment after conception is murder, it's hard not to be struck by the shift. Today's Christian fundamentalists make the ancients and medievals look like contemporary liberals and moral relativists. While the theoconservatives still rely on Aquinas for the teleology of sex, and steadfastly ignore what science has contributed to our understanding of human nature in the past few centuries in that regard, in the question of abortion, they are all in favor of a scientific update. And that update, they argue, demands a more rigorous understanding of the moral enormity of all abortion.

Contemporary natural law philosophers describe legal abortion as an integral part of what they call a "culture of death" that is synonymous with the modern West. The phenomenon of abortion is described bluntly, in Robert George's words, as "the killing of human beings" or "the unjust taking of innocent human life." Pope John Paul II, who put this new and constant human holocaust at the center of his papacy, argued that "procured abortion is the deliberate and direct killing, by whatever means it is carried out, of a human being in the initial phase of his or her existence." His successor underlined this point in his first Christmas address in 2005. Referring to Psalm 138, Benedict XVI declared: "The loving eyes of God look on the human being, considered full and complete at its beginning. It is extremely powerful, the idea in this psalm, that in this 'unformed' embryo God already sees the whole future. In the Lord's book of life, the days that this creature will live and will fill with works during his time on earth are already written."

For the new fundamentalists, the old balance of morality has been abolished in favor of a new absolutism: not only is all abortion at any stage equally immoral, it is always as immoral as killing an adult human being. Today's natural law philosophers do not flinch at this analogy. Indeed, they insist that we not euphemize what is going on. As Pope John Paul II said, "Given such a grave situation, we need now more than ever to have the courage to look the truth in the eye and to *call things by their proper names*, without yielding to convenient compromises or to the temptation of self-deception." What he was saying is clear enough: societies which allow widespread abortion are practicing legal mass murder on an overwhelming scale. This crime is numerically equivalent in America alone to two Holocausts a decade. In America, six million human beings are deliberately and legally murdered, according to the pope, every five years. Hitler was less efficient.

Does this argument stand up? Or was earlier nonfundamentalist Christian theology more persuasive? The core of the argument is that from the moment of conception, a human person exists. Robert George lays out the case with admirable lucidity. His argument does not, this time, require some smuggling in of a moral assertion, unsupported by science. It is based on modern scientific analysis that Aquinas had no way of knowing about:

A human being is conceived when a human sperm containing twenty-three chromosomes fuses with a human egg also containing twenty-three chromosomes (albeit of a different kind) producing a single cell human zygote containing, in the normal case, forty-six chromosomes that are mixed differently from the forty-six chromosomes as found in the mother or father. Unlike the gametes (that is, the sperm and egg), the zygote is genetically unique and distinct from its parents. Biologically, it is a separate organism. It produces, as the gametes do not, specifically human enzymes and proteins. It possesses, as they do not, the active capacity or potency to develop itself into a human embryo, fetus, infant, child, adolescent, and adult.

According to George, this minuscule speck of chromosomes is as much a human being as the reader of this sentence. Sure, it is dependent on the mother—and on surviving the treacherous journey of embryonic and fetal development. But you and I can trace our entire being back to those forty-six chromosomes. That's when we became individuated. That's when we became human persons. Yes, a hair on our heads contains the same forty-six chromosomes. But that hair, left alone or embedded in another human being, cannot develop into a human being by itself. That's what makes that zygote unique—something that never existed before and never will again.

But is it a human person as such—with a soul, and all the moral ramifications of personhood? Catholic teaching remains surprisingly noncommittal on this specific topic. In its most recent formal teaching on the subject, the body tasked with determining doctrine declared the following:

The Magisterium has not expressly committed itself to an affirmation of a philosophical nature [as to the time of ensoulment], but it constantly affirms the moral condemnation of any kind of procured abortion. This teaching has not been changed and is unchangeable. The human being is to be respected and treated as a person from the moment of conception, and therefore from that same moment his or rights as a person must be recognized, among which in the first place is the inviolable right of every innocent human being to life.

So even if we cannot know for sure that the zygote is an actual person with a soul, it is to be treated as such. When you could be in danger of sanctioning murder, better safe than sorry. Natural law philosophers argue that because the genetic code is identical in the zygote to that of a human being at any stage of life, the zygote is inherently indistinguishable in kind from a fully functioning, thinking, breathing, eating adult. George again:

> [T]here comes into being at conception, not a mere clump of human cells, but a distinct, unified, self-integrating organism, which develops itself, truly himself or herself, in accord with its own genetic "blueprint" . . . [N]o outside genetic material is required to enable the zygote to mature into an embryo, the embryo into a fetus, the fetus into an infant, the infant into a child, the child into an adolescent, the adolescent into an adult. What the zygote needs to function as a distinct self-integrating human organism, a human being, it already possesses.

The first thing to note about this argument is how distant it is from our intuition. That isn't necessarily a point against it. Many things that are empirically true escape our intuition. We find it hard to visualize invisible electricity, for example, but it's still there. We find the concept of being weightless very alien to our instinctive sense of things—but outside the earth's gravitational pull, human beings float around like helium balloons. So perhaps it is true that an entity so tiny it is infinitesimal compared to a punctuation mark on this page is still a fully fledged human person.

So let's follow the logic of the argument. If zygotes are full-fledged human beings, then there must surely come a moment—the most miraculous moment in human existence—when life actually comes into existence. This is the clear line that George insists is the only clear and nonarbitrary line in deciding when a human being comes into existence. Science, however, suggests that even this line is somewhat fuzzy. The evolutionary biologist Steven Pinker explains why in his book *The Blank Slate*:

> Just as a microscope reveals that a straight edge is really ragged, research on human reproduction shows that the "moment of conception" is not a moment at all. Sometimes several sperm penetrate the outer membrane of the egg, and it takes time for the egg to eject the extra chromosomes. . . . Even when a single sperm enters, its genes remain separate from those of the egg for a day or more, and it takes yet another day or so for the newly merged genome to control the cell. So the "moment" of conception is in fact a span of twenty-four to forty-eight hours.

The closer you look, the more the bright lines of the fundamentalist blur into the actual reality of human life and development. Another aspect of that reality is the undeniable fact that even the successfully fertilized zygote is more likely to die than to live, without any human intervention at all. In fact, if the zygote is a complete human being, then the female body is necessarily a vast ocean of fathomless grief, a biological graveyard for countless deaths. The statistics vary because of the minuscule phenomenon involved, and the immense difficulty of measuring it with any precision, but the scientific literature estimates that from 30 to as much as 50 percent of zygotes perish of their own accord, failing to develop beyond the most primitive of stages to anything remotely recognizable as a developing baby. Spontaneous abortion of zygotes, blastocysts, and embryos is routine in the human reproductive cycle. It is far more common than successful pregnancy.

Now there is, obviously, a big distinction between the deliberate decision to end the development of a zygotic human being and the fact that nature allows them to perish in huge numbers. But the context should surely make us pause. In his book *It Takes a Family* the pro-life senator Rick Santorum argues that "as a result of abortion for more than thirty years, over a quarter of all children conceived in America never took their first breath." Strictly speaking, he is mistaken. As a result of *all* abortions—spontaneous *and* procured—well over three-quarters of all children conceived in America never take their first breath. If God is so careless with his creation at this stage of its development, one might ask why humans should have much higher standards.

Of course the reason for this constant wave of human death within the human body is plain. We overproduce zygotes in reproduction precisely in order to beat the odds of their limited life spans. If we are to have a good chance of getting one zygote to navigate the fallopian tubes, develop into a blastocyst, attach itself to the womb, and avoid all the subsequent pitfalls of a pregnancy, we need to sacrifice more of them to imminent death. And this, recall, is what science is teaching us. This must, by George's own argument, be the guideline for our moral judgment. On this basis, we all have countless siblings whose lives lasted only a few minutes, hours, days, or weeks, only to perish inside our own mothers' bodies. Most of them die silent, unmarked deaths before a woman is even aware that she is pregnant. The death rate declines as the zygote becomes a blastocyst, the blastocyst becomes an embryo, and the embryo becomes a fetus. But even then, tragedy strikes. Roughly one in ten unborn children perish after a woman realizes she is pregnant, usually between the seventh and twelfth weeks of pregnancy. Miscarriages can occur much later as well. By any measure this is a brutal process of natural culling. If you believe that each one of these doomed zygotes is as valuable and sacred a human person as you or I, the tragedy is so vast it almost defies comprehension.

The new natural law argument, in other words, runs counter to our intuitive sense that human personhood cannot simply be reduced to its most basic genetic component, without any reference to how it functions or what it is. Another way of looking at this is to examine the phenomenon of identical twins, or triplets. Genetic twinning occurs after the formation of a zygote. So was the pretwinned zygote one human person or two? If, by that logic, personhood can be reasonably inferred to occur other than immediately upon conception, then we have no easy solution to the question of when a line can be safely drawn between human life and a human person.

There are, in fact, any number of plausible arguments for the moment when an unborn life becomes an unborn child. There is the development of a more complex cell structure, in the blastocyst. There is the successful implantation into the uterus. There is the moment when a basic brain can be detected; the moment when a rudimentary heart begins to beat; the moment when pain and sensation are inferable; the moment when a fetus can be sustained outside its mother's womb; and so on. All sorts of reasonable criteria could be applied, related to a fetus's bodily functions, cognitive skills, independent viability, and so on. Suddenly Aquinas's forty days seem less arbitrary.

And the truth is that we cannot but project onto unborn life the features of its more developed future self when we attempt to comprehend it. This is true even for those who insist that forty-six chromosomes are a fully formed human being. Here is one of the most passionate opponents of abortion, Senator Rick Santorum, making the case against partial birth abortion:

> When you hear an abortion supporter argue his or her position, nowhere do you hear that a baby's heart can be seen beating at three weeks; that new 4D sonograms show that from twelve weeks, unborn babies can stretch, kick and leap around the womb—well before the mother can feel their movement; from eighteen weeks, they can open their eyes, although most doctors thought eyelids were fused until 26 weeks; from 26 weeks, when partial birth abortions are still performed, they appear to exhibit a whole range of typical baby behavior and moods, including scratching, smiling, crying, hiccupping, and sucking.

Following Santorum's own argument, none of this is logically relevant in the slightest. The death of a seconds-old zygote is just as much a human tragedy as the loss of a child in the fifth month of pregnancy, as Santorum himself sadly experienced in his own life and family. Santorum grieved over his own miscarried child, brought its aborted body back so that his siblings could see the half-formed fetus, named, and buried it. But that lost son had dozens and dozens of siblings whose deaths were just as morally significant, but which Santorum and his wife couldn't possibly have known about. Did he set up a period of grieving after each menstrual cycle? If not, why not?

This does not, of course, logically refute George's and Santorum's fundamentalist case for the grave tragedy of every zygote that never makes it; and the immorality of any person who impedes it. But if "natural law" appeals to our human sense of reason, then we can at the very least say that our intuition strongly resists equating the life of a zygote with that of, say, a two-month-old fetus, or even a premature baby born at the earliest boundaries of viability. We can say that reasonable people, all of whom take human life very seriously, can disagree on the line between human life and human personhood. The basic context for this disagreement is the inexorable fact that in the extremely hazardous journey from conception to birth, death is as much the natural rule as it is the exception. And the more insistently we look for the magic moment when personhood begins, the more elusive it becomes.

Moreover we are more morally repelled by partial birth abortion than by the unknown and unknowable death of a zygote for a reason: one is clearly more recognizable as a human person. What a human being is, as Aquinas understood, is related to what he does. A speck on this page cannot exhibit the characteristics of humanness that even an infant baby can. The very instinct that pushes Santorum into a passionate and persuasive attack on partial birth abortion is an instinct that renders his own fundamentalist case less persuasive. This is not to say that in the absence of certainty, we can end unborn life with abandon. My own view is that Aquinas was roughly right: within around a month after conception, the survivability of an embryo improves dramatically and recognizably human features appear.

Most abortions that we now know of would still be and should be classified as gravely immoral. But it is very hard to reasonably equate premeditated murder with, say, a morning-after pill that merely increases the rate of zygote death. In that distinction lies the journey that nuanced Catholic and Christian moral thinking has traveled in order to become the rigid fundamentalism that so much of it is today. In that distinction lies the difference between reason and fundamentalism.

V

IF THE QUESTION OF WHEN LIFE BEGINS IS A CENTRAL concern for the fundamentalist, so too is the moment when it ends. And just as with the beginning of human life, the chief characteristic of today's theoconservatives is the radically new extremism of their pro-life position. Their view is that there can be no distinction made between the existence of a human being and his or her capacity to act as a human being. A zygote may have no recognizable characteristics that suggest he or she is human, but that doesn't render his or her humanity any less total and sacred. The same goes for, say, a woman in a persistent vegetative state for decades, with no brain waves, no capacity for thinking, feeling, seeing, speaking, or even eating without external support, and no prospect for recovery or survival. The basic, and in principle admirable, principle is what Robert George calls "the equality-in-dignity of all human beings—without regard to age, size, stage of development, or condition of dependency—a principle at the heart of our laws against homicide."

To separate the human being into different parts—a conscious self in contrast to a body that self somehow possesses—is, for George, to violate the integrity of the human being herself. There is not, on the one hand, "biological life" and then, on the other, in philosopher Ronald Dworkin's words, "exercisable abilities, especially for rational control of one's life, in virtue of which people can give the shape and significance they wish to their lives." There is simply one indivisible, continuous human being. The British Linacre Centre put the argument this way:

> One's living body is instrinsic, not merely instrumental to one's personal life. Each of us has a human life (not a vegetable life plus an animal life plus a human life); when it is flourishing that life includes all one's vital functions, including speech, deliberation, and choice; when gravely impaired, it lacks some of those functions without ceasing to be the life of the person so impaired.

The theoconservatives have an obvious but vital insight here. If we disaggregate human life itself into a mere series of functions, and describe a human being only as someone with the full range of those functions, or even a minimal level of those functions, then we are in danger of sanctioning all sorts of eugenic or social reasons for the killing of "subhuman" human beings. If a human being is defined by rational self-consciousness, then patients with Alzheimer's might be deemed subhuman. If a human being is defined by a full capacity for physical flourishing, then those with extreme physical disabilities might be regarded as somehow less than human. If we define being human as being able to fear death or remember the past or make plans for the future, or experience love, then vast numbers of human beings, impaired by physical or mental disability could be put beyond the boundary of civilization. And this not only violates the dignity of each, it opens up a terrifying prospect of legalized murder of some of the most vulnerable people in our world.

So let's take the natural law theory seriously, since it certainly offers a moving and powerful defense of all human dignity. Given the capacity for technology to extend life in ways our predecessors never dreamed of, how do we specifically define the moment of death? And what measures are morally necessary to keep someone alive, even after all hope for recovery has been lost?

Again it's worth recalling the insights and teaching of the earliest natural law philosophers. They did not have the technological capacities we have, but the underlying principles were the same. The basic argument was that we should do all we can to protect and extend human life, but that there comes a point at which extending it by extreme and burdensome means is unnecessary. The esteemed Jesuit philosopher John Paris, a bioethics professor at Boston College, summed up the centuries-old doctrine for *Newsweek* in 2005:

> The church doctrine, and it's been consistent for 400 years, is that one is not morally obliged to undergo any intervention [to prolong one's life]. And, of course, 400 years ago they weren't talking about high technology. Here's the example one of the moralists of the 16th century gave: if you could sustain your life with partridge eggs, which were very expensive and exotic, would you be obliged to do so? The answer is no, they're too expensive. They're too rare. You can't get them. They would be too heavy an obligation to put on people.

The Church once made a distinction between ordinary measures to extend and preserve life and extraordinary ones. The ordinary ones might be treating a cancer aggressively; or even trying a risky new procedure to deal with a particular disease that had outrun all previous attempts to restrain it. But if a treatment might prove a terrible burden to an already terminally sick person, there was no ethical obligation to take every measure to impose it. If a sick medieval monk believed that he could get a cure at a shrine fifty miles away, he might be obliged to give it a try. But if the shrine were five hundred miles away, he would not be abetting a "culture of death" in resigning himself to his fate.

This distinction can be difficult to pin down with any certainty in any given case. That's where prudence comes in—the wisdom to determine a practical course of action in a specific place and time. I recall being present when one of my closest friends died of AIDS at the age of thirty. His immune system had essentially collapsed and his lungs had filled with liquid; his vital signs were in sudden free fall; at one point, his pulse simply stopped. Should we jolt him back for a few more hours of life? If we didn't, were we killing him? No one outside any particular, unique case can say for sure. My friend had asked us specifically not to go to extreme measures to extend his life artificially. We complied. Even at his young age, he had already reconciled himself to death. As soon as his last close friend had made it to the hospital to say good-bye, he let go. And we let him let go. There is a balance here between a respect for life and a fetishization of it.

A former boyfriend of mine had a mother who suffered a massive stroke and was subsequently deemed in a persistent vegetative state. She was taken into care, and in any other age, would have died relatively soon thereafter. In the modern world, she lived for another decade and a half, unable to speak or see or think or move on her own. She suffered the stroke when my ex was a teenager. He has a heartbreaking photograph of his mother at his college graduation: propped up in a wheelchair, sealed off from any understanding of what was going on around her, her face positioned by others to stare emptily at the camera, but physically present for a moment in her life that would normally be viewed as a milestone. Was she really there? If the George argument was correct, she was. And in outward appearance, she looked recognizably human. All I can say is that my ex took a few years to absorb the fact that his mother had actually ceased to exist in any meaningful way. But she stayed alive. When a bacterial infection finally overcame her, his sister called not to tell him that their mother had died, but with some less definitive words: "It's over."

I do not conclude from that experience that his mother should have been killed, or even allowed to die through benign neglect. I merely conclude that the boundaries between life and death can be far less sharp than we would like them to be; and as medical science advances into what previous generations would have deemed miraculous territory, those boundaries can become less and less clear. Pinker again makes the point very elegantly: "Most people do not depart this world in a puff of smoke but suffer a gradual and uneven breakdown of the various parts of the brain and body. Many kinds and degrees of existence lie between the living and the dead, and that will become more true as medical technology improves."

There are also kinds and degrees of treatments that can stave off death or extend life. At what point do these become extraordinary? Again context matters. My grandfather, in his mid-nineties, finally collapsed and entered the hospital with pneumonia. I saw him in his last days, coiled into a ball, smaller than I had ever seen him, almost as fetal in his end as in his beginning. Should we have let the infection take its course? We thought not. He was given a massive dose of antibiotics, and as someone who had largely refused medical care for most of his life, he had a dramatic response. For a few days, he regained energy. And then a very short time later, he died. I think we made the right call. But it was equally clear that he was ready to die, and beyond ordinary care, it would have been inappropriate to extend his life indefinitely.

In other words: there's a balance of goods here. Life as such can never be an absolute good regardless. It has to be balanced against the quality of life experienced, the chances for recovery, the likelihood of death, and the medical options available. Letting someone die is not the same as killing someone. But that principle is precisely at the core of the theoconservatives' challenge. It is precisely that distinction that they wish to abolish.

The Terri Schiavo case provided a dramatic example of what theoconservatism really means. Schiavo had been in a bed in a persistent vegetative state, unable to perform even basic functions of feeding herself, for fifteen years. Her brain had shrunk to near nonexistence. She could not see, speak, think, move, eat, or drink. In a previous century, she would have died a decade and a half before she did. The critical moral question in her very bitter end was whether an artificial feeding tube that provided a sophisticated blend of nutrients and hydration to keep her body intact was an ordinary or extraordinary measure to maintain what remained of her life. The natural law theologians who deemed a partridge egg too expensive to save an individual's life in the middle ages would not have had a dilemma. They would have seen such a practice as extreme. The huge expense—along with the profound bodily pain and discomfort caused by a decade and a half of being bedridden—would have ruled out indefinite care.

But the new fundamentalism—embodied by the "culture of

life" endorsed by Pope John Paul II and his closest adviser and successor, Pope Benedict XVI—decreed otherwise. In a 2004 speech at the Vatican, John Paul had specifically included the indefinite use of a feeding tube as a merely "ordinary measure" for sustaining life. "The administration of water and food, even when provided by artificial means, always represents a natural means of preserving life," even to patients in a vegetative state with no hope of recovery, he said. Someone in a vegetative state "has the right to basic health care (nutrition, hydration, cleanliness, warmth, etc.) and to the prevention of complications related to his confinement to bed." It was morally obligatory.

Feeding tubes have only been in use for a quarter of a century; the medical care needed to flush them, maintain them, monitor intake, and insert and replace them is considerable. They carry high risk of infection—often repeated infection in patients with long-term needs. The following is a list of ingredients for one of the most common formulas: "Water, maltodextrin, soy oil, calcium caseinate, sodium caseinate, MCT oil, soy protein isolate, and less than 0.5%: soy lecithin, carrageenan, vitamin A palmitate, vitamin D_3, vitamin E acetate, sodium ascorbate, folic acid, thiamin hydrochloride, riboflavin, niacinamide, vitamin B_6 hydrochloride, vitamin B_{12}, biotin, calcium pantothenate, vitamin K_1, choline chloride, calcium citrate, calcium carbonate, potassium iodide, ferrous sulfate, magnesium phosphate, magnesium chloride, cupric sulfate, zinc sulfate, manganese sulfate, potassium citrate, potassium chloride, sodium citrate, sodium selenite, chromic chloride, sodium molybdate." All this is blended into a highly liquid substance to prevent clogging of the narrow tube. If this is the standard for ordinary care, then the expense of keeping every terminally ill patient permanently on feeding tubes would bankrupt even the wealthiest country on earth.

To write that last sentence is not, I would argue, to abet in a "culture of death." When it is technology alone that has created exponentially receding new boundaries for the sustenance of barely functioning human life, when people born in previous decades, let alone centuries, would have never even been able to contemplate these dilemmas, the saner and more rational approach is a prudential balancing of goods. The positing of life itself—any life, however unrecognizable as a flourishing human life—as an absolute good to be protected at any price is also a very strange doctrine for Christians. John Paris again:

> Richard McCormick, who was the great Catholic moral theologian of the last 25 years, wrote a brilliant article in the *Journal of the American Medical Association* in 1974 called "To Save or Let Die." He said there are two great heresies in our age (and heresy is a strong word in theology—these are false doctrines). One is that life is an absolute good and the other is that death is an absolute evil. We believe that life was created and is a good, but a limited good. Therefore the obligation to sustain it is a limited one. The parameters that mark off those limits are your capacities to function as a human.

This is the voice of Christian moral reasoning. It is not the voice of theoconservatism. The latter is extreme and unpersuasive even within the boundaries of its own logic. But when it comes to dictating the boundaries of our political life, its impact is even more unsettling. And in the new millennium, it got a chance to put its own extreme beliefs into practice.

CHAPTER 4

The Bush Crucible

"We have a responsibility that when somebody hurts,
government has got to move."
—PRESIDENT GEORGE W. BUSH, *September 2003*

I

FOR THE TRUE FUNDAMENTALIST, POLITICAL LIFE IS EX-
tremely simple. Fundamentalist politics demands that the truths
handed down by revelation or "nature" be applied consistently to
govern all citizens. How could they not be? For the fundamentalist, there is no secular truth independent of religious truth;
and there is no greater imperative than saving souls. If there is
an inevitable disjunction between the demands of Heaven and
the brokenness of Earth, the role of politics is to narrow that
gap as far as possible. It is to conjoin ultimate meaning with a
monopoly of force.

The ultimate face of fundamentalist politics can be seen
throughout history. From the Catholic Spain of the Inquisition to the Puritan dictatorship of England's Oliver Cromwell,
the same fanaticism and conflation of dogma and law crops up
again and again. In the hideous tyrannies of Afghanistan under
the Taliban or of Iran after the fall of the shah, you can see
what happens when religious truth merges with modern political power. The secular equivalents—regimes constructed to
propagate and disseminate an inerrant doctrine and to find and

punish all error—are the dictatorships of Hitler, Stalin, and, at the most extreme, Mao and Pol Pot. Sometimes the doctrine is entirely modern—such as the North Korean doctrine of Juche, concocted by the mass murderers, Kim Il Sung and his son, Kim Jong Il. These vicious regimes tend not to last too long. Built on the banishment of all human freedom, and resting on religious or pseudo-religious doctrines that cannot brook any compromise with existing reality, they are bound to fail in the end. But while they last, and they can last decades, they are the vehicles for some of the worst brutality, evil, and criminality that the world has ever known.

To equate the astonishing rise of evangelical and Catholic fundamentalism in the contemporary West with these monstrous regimes would be absurd. There are a few fringe groups in America—the Christian Reconstructionists, for example— who would like to replace the United States constitution with biblical law, a Christian version of sharia. But they are marginal, extremist, and largely disowned by the fundamentalist mainstream. Evangelical and Catholic fundamentalists have largely engaged in America in completely legitimate and democratic activity: voting, organizing, campaigning, broadcasting, persuading. Even where they disagree with the Supreme Court's interpretation of the Constitution, they do not question that Constitution's legitimacy (although a few have indeed walked to the brink of declaring the United States an illegitimate "regime" because of the court's rulings). They constantly use religious language to defend their political positions—but so did Martin Luther King Jr. and Abraham Lincoln. The political methods of the new fundamentalists are overwhelmingly democratic ones. They are not, properly speaking, theocrats, if by that we mean explicitly religious rule by religious figures with no constitutional safeguards for minorities or individuals. A person who believes that society should be governed only by laws consistent with his religious faith is not a theocrat if he merely tries to persuade majorities of his case, and restricts himself to constitutional, legal, and nonviolent activity.

But this much can nevertheless be said of the new American fundamentalists: they deny the possibility of a government that is neutral between differing views of what the meaning of life truly is. They reject the whole idea of the law as a way to create a neutral public space, to mediate between competing visions of the good, to provide an umpire for a game between competing visions of what is moral, right, or true. They reject, in short, the entire premise of secular democracy: that religion should be restricted to the private sphere, and the law should be as indifferent as possible to the substantive claims of various impassioned groups of true believers.

They are extremely clear on this point. Robert George argues that "there can be no legitimate claim for secularism to be a 'neutral' doctrine that deserves privileged status as the national public philosophy." It's worth insisting here on the proper meaning of secularism. It is not antireligious, as is now often claimed. Definitionally secularism merely argues that public institutions and public law be separated from religious dogma or diktats. A secular society can be one in which large majorities of people have deep religious faith, but in which politics deals with laws that are, as far as possible, indifferent to the religious convictions of citizens, and clearly separated from them.

Some evangelical Christians have thrived under secularism. Think of the astonishing career of someone like Billy Graham, a man who sought to bring millions to a fundamentalist faith but who didn't construct a political movement as such. Or think of the work of a group like the Salvation Army that, in a secular society, channels fundamentalist faith into social action and helps so many people in real need. It's possible for evangelicalism to coexist with secularism—and, indeed, for much of American history, that has been the case. But it is no longer the predominant view on what we might call the "religious right," or within the Republican Party as a whole.

In fact the key premise of secular neutrality is precisely what the Christianist right now disagrees with. Senator Rick Santorum affirms, "I don't want a government that is neutral between virtue and vice." Elsewhere Santorum writes that defenders of secularism "are trying to instill a different moral vision—one that elevates the self, the arbitrary individual good, above all else. And frankly, this moral vision amounts to nothing less than a new religion, a polytheistic one in which each individual is to be his own god to be worshipped." For Santorum, the alternative to a politics responsive to one God is a politics responsive to many, with no mechanism to distinguish between them.

I should mention in passing that, in this conviction, Santorum has philosophical allies on the left. Left-fundamentalists also discount the whole idea of government neutrality and want to use government power and the law to insist on their own values: racial justice (by affirmative action), enforced tolerance (through restrictions on free speech and hate crime laws), direct public funding of deeply controversial areas like abortion, and the use of public schools to inculcate the dogmas of multiculturalism in children. The two overlap in some areas. Restricting pornography, for example, is a top priority for the religious right and the fundamentalist left. The point here, however, is their mutual disdain for the idea of public neutrality. For the fundamentalist left, it's a mask designed to obscure systematic oppression of various kinds; for the fundamentalist right, it's a sham created to obscure a "secular humanist" agenda and even, as Santorum would have it, polytheism.

Classical liberals and secular conservatives differ. They cling to the notion that government can try to be above the fray, that it can aspire to be the mediator for very different people who have to live alongside one another with radically divergent ideas of what is good or true. These limited-government liberals and conservatives believe, as a critical part of this notion of politics, that there is a clear distinction between what is public and what is private. The law can allow for different moral choices, they argue, without privileging one over the other. So a law that permits abortion merely allows some women to choose it and others to refuse it, according to their own views of what is right and wrong. And a law that allows for legal pornography simply lets individuals make their own decisions, and takes no stand on the underlying issue itself. A law that allows gay couples to marry does not forbid straight couples from marrying. The law, in this sense, is indifferent to what any couple or person might choose. It just grants them the right to choose it, and provides the mechanisms to defend the choice.

Fundamentalists reject this idea. For them, neutrality is a sham. In any law, they argue, someone's values are asserted. A law that permits abortion is a law that says abortion is morally permissible. It is de facto pro-abortion, not pro-choice. A law that allows pornography or legal contraception is not neutral with regard to these alleged moral evils. It abets them, approves them, legitimizes them—and all the immoral behavior they imply.

How? Because law teaches people culture. By merely permitting abortion, the law sends a strong message that abortion is okay. This very teaching can pollute people's minds, render them more susceptible to the evils involved, and foster an immoral society. Remember the insistence of the young fundamentalist in the last chapter: There can be no gray here. There can only be encouragement or discouragement of certain activities. Here's Santorum attacking a Supreme Court ruling that spoke of the meaning of freedom:

> I reiterate what the Court said in the case of *Planned Parenthood v. Casey*: "At the heart of liberty is the right to define one's own concept of existence, of meaning, of the universe, and of the mystery of human life." Let me put that in plain English: you make your own laws; you are a god. The world liberals see is one in which each of us is an island entire unto ourselves supported by the village elders, doing what pleases us, as the U2 song says, "as long as nobody gets hurt." Except that everybody gets hurt.

What people do privately, according to Santorum, affects the public atmosphere, the broader culture, and that culture in turn affects the choices people make. Virtue is at stake. Morality is threatened under the guise of "freedom." There is no way, the fundamentalists argue, to separate the public from the private so neatly and cleanly. As George explains: "Public morals are affected, for good or ill, by the activities of private (in the sense of 'non-governmental') parties, and such parties have obligations in respect to them."

Take the question of prostitution. Should it be legal? A classical liberal or secular conservative might argue, Why not? The transaction is between two adult parties; there is mutual consent; the conduct occurs in private; if public solicitation is avoided, the public space is not directly affected. George begs to differ:

The public has an interest in men not engaging prostitutes; for when they do, they damage their own characters; they render themselves less solid and reliable as husbands and fathers; they weaken their marriages and their ability to enter into good marriages and authentically model for others (including their children) the virtue of chastity on which the integrity of marriages and of marriage as an institution in any given society depends; they set bad examples for others.

Immoral decisions, in other words, are like environmental pollutants. They cannot be given the protection of the law because they affect others indirectly. Society as a whole has an obligation to protect its moral environment as much as its physical environment. And this argument must logically extend far beyond mere commercial transactions like the purchasing of sex. It must also logically include those noncommercial private immoral acts that also damage an individual's moral character, and so set bad examples, and contribute to the breakdown of society as a whole. George is admirably frank about this. He sees some prudential, practical reasons why the government cannot outlaw all private vice. It would require the government to be able to police homes and bedrooms, and thus impose too great a burden on their security and sense of liberty, and be astronomically expensive. The sheer number of cops required to prevent, say, the unnatural evil of masturbation (as wrong, in George's view, as adultery, contraception, or lesbian sex) would be legion. But George sees no reason *in principle* to object to such an idea:

It is, of course, true that prostitutes and their pimps are inviting and doing business with "the public" in a way that ordinary fornicators and adulterers are not. And I can certainly see how this distinction could be relevant to the prudential reasoning of legislators considering enactment or repeal of legal prohibitions of noncommercial sexual vice. What I cannot see, however, is the ground for claiming that *a strict principle of justice* excludes the criminal prohibition of noncommercial sexual vice.

This is one reason fundamentalists were so outraged by the Supreme Court's decision to declare unconstitutional the arrest of two Texas men in their own bedroom for the crime of non-procreative sex. Supreme Court Justice Antonin Scalia, in dissent in *Lawrence v. Texas*, argued that the notion of inviolable, constitutionally protected privacy for adult consensual sex would render all sorts of moral prohibitions impossible: "State laws against bigamy, same-sex marriage, adult incest, prostitution, masturbation, adultery, fornication, bestiality, and obscenity are likewise sustainable only in light of *Bowers*'s validation of laws based on moral choices. Every single one of these laws is called into question by today's decision; the Court makes no effort to cabin the scope of its decision to exclude them from its holding. . . . This effectively decrees the end of all morals legislation."

By "morals legislation," Scalia means the ability of the government to criminalize private, adult, consensual activity, like masturbation or sex outside marriage. For Scalia, there is no constitutional freedom to have sex in one's own bedroom, to consume pornography, or even to masturbate. The government has a direct interest in monitoring and restraining such behaviors to uphold public morals.

What happens to the idea of freedom here? Something similar to what happens to the idea of conscience in the thought of Pope Benedict XVI. It is redefined to mean something more than a little different than the conventional meaning. Santorum tackles the point head-on: "At first, the liberal vision may sound attractive—because freedom is attractive. The only problem is that it is a false vision, because nature is nature, and the freedom to choose against the natural law is not really freedom at all." The freedom to determine one's own way of life, to make up one's own mind about the meaning "of the universe, and of the mystery of human life," is not really freedom, according to Santorum. And notice the necessary logical reference to "natural law." Without that philosophy as sketched in the previous chapter, Santorum would himself have no way to decide what is or is not moral. Without the pope's inerrant interpretation of that law, he would be equally adrift.

What is freedom then? Santorum tells us:

> Properly defined, *liberty* is freedom coupled with responsibility to something bigger or higher than the self. It is the pursuit of our dreams with an eye toward the common good. Liberty is the dual activity of lifting our eyes to the heavens while at the same time extending our hearts and our hands to our neighbor. In other words, our founders' understanding of liberty ordered the individual toward a higher good, defined in part by our Judeo-Christian roots.

Santorum here stumbles across one major obstacle to his view of America and of American freedom. It's worth taking a moment to consider it. The Founders, to Santorum's dismay, did not write a Constitution dedicated to the inculcation of virtue. In fact what is stunning about the American Declaration of Independence and subsequent Constitution is how morality and virtue are all but absent as a primary concern. The tripartite goal of the American founding was "life, liberty and the pursuit of happiness." They did not write "the pursuit of virtue" or the "pursuit of morality." Moreover the First Amendment guaranteed that religious freedom—and therefore religious diversity—would be integral to the new republic. At its very inception, in stark contrast to both the French Revolution and to the English establishment of a state religion, Americans insisted on freedom first. Any single version of religion or morality would be up to the choice of individual citizens in the privacy of their own consciences. And the founders—from Madison to Jefferson—were clearly skeptics, open-minded deists, or very different Christians than those who call themselves fundamentalists today. In their Constitutional deliberations, they also appear far from obsessed with religious doctrine. They were more interested in thinking about the failures of Roman and Greek democracy than in the intricacies of Thomas Aquinas.

How do the new fundamentalists deal then with America's deist founding, its original fear of established religion? How do they grapple, say, with Madison's insistence that an "alliance or coalition between Government and religion . . . cannot . . . be too carefully guarded against. . . . Every new and successful example therefore of a perfect separation between ecclesiastical and civil matters is of importance"? How do they account for the fact that the first presidents avoided public prayers? How do they reconcile themselves to Jefferson's insistence on the private nature of religious faith: "Our particular principles of religion are a subject of accountability to God alone. I inquire after no man's, and trouble none with mine." Many of the founding fathers disputed the divinity of Christ. John Adams viewed the founding of the United States thus:

> Although the detail of the formation of the American governments is at present little known or regarded either in Europe or in America, it may hereafter become an object of curiosity.

It will never be pretended that any persons employed in that service had interviews with the gods, or were in any degree under the influence of Heaven, more than those at work upon ships or houses, or laboring in merchandise or agriculture; it will forever be acknowledged that these governments were contrived merely by the use of reason and the senses.

These are not the words of fundamentalists, let alone of people who believed that the government should be directly involved in the religious or moral life of its own citizens. Compare them, for example, with the words of a leading Republican senator, Sam Brownback of Kansas. According to a firsthand account in *Rolling Stone* magazine, Brownback has a very particular understanding of his political role:

> [Brownback] tells a story about a chaplain who challenged a group of senators to reconsider their conception of democracy. "How many constituents do you have?" the chaplain asked. The senators answered: 4 million, 9 million, 12 million. "May I suggest," the chaplain replied, "that you have only one constituent?" Brownback pauses. That moment, he declares, changed his life. "This"—being senator, running for president, waving the flag of a Christian nation—"is about serving one constituent." He raises a hand and points above him.

How to reconcile this view of politics with the skeptical deism of the founders? Today's fundamentalists have an answer to this apparent conundrum. Rick Santorum expresses the revisionist view succinctly enough: "It was because America's founders expected such *public* support for religion and morality—religion and the family, the seedbeds of virtue—to be natural, even inevitable, in a democracy that they did not think it necessary to formally incorporate provisions concerning virtue into the Constitution."

This is surely stretching the point. One thing you can say about the founders of the United States is that their eyes were fixed on the long view. They were by no means captives of the cultural milieu in which they found themselves. They openly considered and discussed many cultures and polities throughout the ages while they deliberated on a constitution for a "*novus ordo seclorum.*" They assumed nothing. And their decision to rest America on freedom rather than virtue was surely not made in a fit of absentmindedness.

To be sure, there is no doubt that the founders—and indeed many liberal theorists—viewed personal and civic virtue as an important component of a flourishing constitutional republic. Without trust in one's fellows, without honesty in public office, without public-spiritedness or personal sacrifice for the greater good, free societies could succumb to factionalism, corruption, social decay. But the founders were just as adamant that the in-culcation of virtue was primarily the task of the private sphere, of individuals figuring out their own moral systems or religious faiths, of schools and parents teaching children the virtues they wanted to pass on, of churches free of any government med-dling, inspiring and helping people to lead good lives. But they were insistent that the government should not prefer one church to another, or one faith to another. They knew where that ulti-mately led: to the religious warfare that had destroyed Europe in the seventeenth century. They were content to let Americans in their private capacity discover their own truth, forge their own path, pursue their own happiness. And that included not just every variety of Christian, but also atheists and those who disdained religious belief as a whole.

The man who actually came up with the name of the new nation, the United States of America, was, after all, an atheist. Thomas Paine loathed religion of all kinds: "I do not believe in the creed professed by the Jewish church, by the Roman church, by the Greek church, by the Turkish church, by the Protestant church, nor by any church that I know of. . . . Each of those churches accuse the other of unbelief; and for my own part, I disbelieve them all." Not exactly the sentiments of George W. Bush, who has often stated that faith itself—regardless of its object—is somehow better than unfaith. The founder of the country, George Washington, was particularly concerned about the impact on civil life of religious and doctrinal disputes. He wrote in a letter: "Religious controversies are always productive of more acrimony and irreconcilable hatreds than those which spring from any other cause. I had hoped that liberal and enlightened thought would have reconciled the Christians so that their religious fights would not endanger the peace of Society." Tell that to the architects of Republican hegemony, Karl Rove and Tom DeLay.

Jefferson believed that the truth was accessible, and virtue reachable, without government intervention at all: "It is error alone which needs the support of government. Truth can stand by itself." He insisted that "the day will come when the mystical generation of Jesus, by the Supreme Being as his father, in the womb of a virgin, will be classed with the fable of the generation of Minerva in the brain of Jupiter." No contemporary American politician could say such a thing without destroying his career in public life.

In 1797 the U.S. Senate unanimously approved the Treaty of Tripoli, an attempt to deal with Muslim piracy and terrorism in the Mediterranean. One of its clauses reads: "As the Government of the United States of America is not, in any sense,

founded on the Christian religion; as it has in itself no character of enmity against the laws, religion, or tranquillity, of Mus-
sulmen; and, as the said States never entered into any war, or
act of hostility against any Mahometan nation, it is declared by
the parties, that no pretext arising from religious opinions, shall
ever produce an interruption of the harmony existing between
the two countries." It is hard to think of a leading contemporary
Republican insisting that American government "is not, in any
sense, founded on the Christian religion." In the early republic,
not a single senator dissented.

For today's fundamentalists, the skepticism, deism, or outright atheism of the founders of America are expressions of
moral nihilism and relativism. Just as the number of Americans
directly involved in religious organizations has grown to levels
unheard of during the late eighteenth century, so too the theory
of American government has shifted to one in which the state
plays a central and often explicitly moral and religious role. For
the new fundamentalists, government has a clear and important role. It is to promote virtue. Whose virtue? The virtues that
are prescribed by what they call the Judeo-Christian tradition.
What are those virtues? If you read the Gospels, you can find
a dizzying array of possibilities: the duty to give to the poor,
to leave one's own family for the service of Christ, to observe
the Sabbath, to refrain from adultery, to visit the sick and the
imprisoned, and so on. The role of government therefore necessarily becomes quite expansive. One leading natural law fundamentalist, John Finnis, is quite clear about the scope of his big
government moralizing conservatism:

> [T]he political community's rationale requires that its pub-
> lic managing structure, the state, should deliberately and

publicly identify, encourage, facilitate, and support the truly worthwhile (including moral virtue), should deliberately and publicly identify, discourage and hinder the harmful and evil, and should by its criminal prohibitions and sanctions (as well as its other laws and policies) assist people with parental responsibilities to educate children and young people in virtue and to discourage their vices.

That's some political mandate. It requires that the government be intimately involved in every aspect of citizens' lives, including the education of their own children, the support of their parents, the morals they uphold, and the decisions they take. There may be prudential limits on how intrusive the government might be in ensuring the cultivation of virtue, but I see no clear line over which the government cannot in principle cross in inculcating human virtue. Laws could conceivably be passed to demand that a certain proportion of someone's income be devoted to charity; government could teach couples how to have better marriages; it could bar pornography; it could recriminalize sodomy and adultery; it could ban contraceptives; it could and should provide mandatory morals education in public schools. And, predictably, efforts have been made—with varying success—on all these fronts in the last decade or so.

II

THE MOVE TOWARD FUNDAMENTALIST POLITICS BEGAN IN earnest in the 1990s, as ascendant Republicans, in the wake of the cold war, focused their attention on domestic issues, primarily moral ones. The defeat of the Soviet Union removed the anticommunist glue that had done so much to keep the conservative coalition together. Moralism was to become the new fixative. In the presidency of Bill Clinton, the fundamentalists saw an opening. Although Clinton moved the Democratic Party sharply toward the right, although he balanced the budget, reformed welfare, restrained government spending, signed the death warrant for a retarded man, and liberalized trade, he was still, to the fundamentalist right, an unrepentant leftist. To sustain this somewhat unconvincing account of a man who privately called himself an Eisenhower Republican, the Republican fundamentalists focused on his private life and his long-standing history of sexual harassment, impropriety, and adultery. For these Republicans, the private-public distinction was all but meaningless. A president had only a right, in the words of special prosecutor Kenneth Starr, to a "private *family* life." If a president engaged in private activity that did not conform to such a moral family life, then he was fair game. Given what we know of the fundamentalist mind-set, this was surely no surprise.

The former president was, of course, responsible for his own expansive batch of lies, mistakes, and abuses of power. He perjured himself, committed sexual harassment, and was foolish enough to provoke his opponents to greater and greater levels of outrage. He was not an innocent victim of a phony witch-hunt. He was a largely guilty victim of an excessive witch-hunt. And what's significant here in retrospect is not so much Clinton's own perfidy but how this era ushered in a conservatism indistinguishable in many respects from the fundamentalist view of politics I have just outlined. In the 1980s Reaganite and Thatcherite conservatism was intent fundamentally on rolling back the economic power of the regulatory state. It was determined to cut taxes, control spending, weaken unions, free trade, and defeat the leftist behemoth of the Soviet Union and its satrapies and allies. Yes, both Thatcher and Reagan had some socially conservative views. But these never supplied the organizing principle or ideological mandate on which they governed. They spoke of freedom far more often than of virtue.

Reagan's vision was essentially a sunny, optimistic one. He rarely went to church, married twice, had many social friends in liberal Hollywood, and, when disposed to tell a morality tale, would prefer an anecdote about someone who overcame adversity rather than a cautionary fable about hellfire and damnation. Thatcher had no religious base in a largely secular country. Both saw government's role as getting out of the way of people's lives, money, and families. President Reagan's most famous quote from his 1981 Inaugural Address was the following: "In this present crisis, government is not the solution to our problem. Government is the problem." President George W. Bush explained a markedly different philosophy in September 2003: "We have a responsibility that when somebody hurts, government has got to move."

Under the first president Bush, the pro-freedom, leave-us-alone coalition of the Reagan era was replaced by moderate realist pragmatism. The first Bush raised taxes and controlled spending, thereby creating the fiscal foundation for the enormous burst of wealth-creation that took place in the 1990s. He managed the collapse of the Soviet Union by emphasizing prudence, realism, and global stability. In the Gulf War, he engaged in only very limited military action, with as large an international coalition as he could muster, under unequivocal United Nations support. This modest pragmatism and discomfort with social issues made George H. W. Bush a marked man within the burgeoning ranks of his increasingly fundamentalist party.

At the time the theorists of the new conservatism were entirely candid about their goals and philosophy. In the Monica Lewinsky saga that consumed American politics for much of Bill Clinton's second term, they saw an entire worldview in the balance. "What's at stake in the Lewinsky scandal," neoconservative writer David Frum wrote in a 1998 issue of *The Weekly Standard*, "is not the right to privacy, but the central dogma of the baby boomers: the belief that sex, so long as it's consensual, ought never to be subject to moral scrutiny at all." The Starr Report reflected this concern. Rather than merely detailing President Clinton's legal missteps, the report pored over the moral iniquities of the man in the Oval Office. There was the sudden prim reminder that "the President's wife" was out of the country during one of Clinton's hallway trysts. There was the superfluously wounding inference that the president was considering leaving his wife after his second term, and the inclusion of the date for one of the president's liaisons: Easter Sunday. The president was on trial for his immorality as much as his illegality. In the fundamentalist mind-set, the two are intertwined.

No conservative thinker did more to advance this new moralism than William Kristol, best known for his urbane appearances on the Fox News Channel and about as close as Washington has to a dean of intellectual conservatism. And no journal did more to propagate, defend, and advance this version of conservatism than the magazine Kristol still presides over, *The Weekly Standard*, launched in 1995 by Rupert Murdoch. In May 1998—at a time when it seemed the Clinton-Lewinsky scandal might dissipate—Kristol urged Republican Congressional candidates to forget other issues in the fall and campaign solely on the issue of the President's morals. "If [Republicans] do that," he argued, "they will win big in November. And their victory will be more than a rejection of Clinton. It will be a rejection of Clintonism—a rejection of defining the presidency, and our public morality, down."

The Weekly Standard was relentless in presenting the scandal as a moral crisis for the nation. Thanks to the President's affair with Lewinsky, *The Standard*'s writers were finally able to see unreservedly in Clinton what they had desperately tried to see in him from the start, but which Clinton's own small-c conservatism had blurred: the apotheosis of the 1960s. The Clinton White House, in the liberated words of Peter Collier in *The Standard,* was "a place where denatured New Left politics meets denatured New Age therapeutics." In February 1998 *The Standard* put on its cover a cartoon of Clinton-as-satyr on the White House lawn grappling with a nude Paula Jones and a nude Monica Lewinsky, surrounded by other naked women in bushes and on a swing, with the one-word headline, "Yow!" Almost one out of two subsequent covers in 1998 focused on the Lewinsky affair. One of the few breaks from Lewinsky coverage was a September cover article on Clinton's alleged genesis. "1968: A Revolting Generation Thirty Years On," the headline blared.

The only issue to rival Lewinsky for prominence among conservative intellectuals in the late 1990s was homosexuality. To anyone familiar with the theorists of natural law, the connection was obvious. Both Clinton's trysts and all gay relationships violated natural law, which it was and is the duty of civil law to uphold. And so mainstream conservatism allied itself with the "reparative therapy" movement to "cure" homosexuals by a strange admixture of Freudian therapy and religious conversion. In June 1997 Kristol gave the concluding address at a Washington conservative conference dedicated, as its brochure put it, to exposing homosexuality as "the disease that it is."

Kristol shared the podium with a variety of clergy members and therapists who advocated a spiritual and psychoanalytic "cure" for homosexuals. One speaker, a priest, described homosexuality as "a way of life that is marked by compulsion, loneliness, depression and disease," comprising a "history-limiting horizon of a sterile worldview divorced from the promise and peril of successor generations." Another speaker decried legal contraception and abortion as the "homosexualization of heterosexual sex," and bemoaned that nonprocreative trends among white Europeans were leading to "race death." In 1996 David Frum called in *The Weekly Standard* for the "reaffirmation by states of a sodomy law" to intimidate homosexuals from campaigning for marriage rights. These were themes that would become completely orthodox among Republicans in the new millennium.

But there was one issue above all others at the center of this new conservatism: abortion. Here is Kristol, explaining the centrality of abortion to the new conservatism almost a decade ago:

> Republicans talk a lot about being a majority party, about becoming a governing party, about shaping a conservative future. Roe and abortion are the test. For if Republicans are incapable of grappling with this moral and political challenge; if they cannot earn a mandate to overturn Roe and move toward a post-abortion America, then in truth, there will be no conservative future. Other issues are important, to be sure, and a governing party will have to show leadership on those issues as well. But Roe is central.

That is why, eight years later, Kristol and many others did the unthinkable and attacked their own president's Supreme Court nominee, Harriet Miers, and forced her to withdraw. With *Roe v. Wade* in the balance, Kristol and his fellow fundamentalist conservatives were not prepared to risk backing someone without a clear record of opposing that Supreme Court decision. In the February 1997 issue of the neoconservative political monthly *Commentary*, Kristol explained why he was so insistent on the issue. "The truth is," he wrote,

> that abortion is today the bloody crossroads of American politics. It is where judicial liberation (from the Constitution), sexual liberation (from traditional mores) and women's liberation (from natural distinctions) come together. It is the focal point for liberalism's simultaneous assault on self-government, morals and nature. So, challenging the judicially imposed regime of abortion-on-demand is key to a conservative reformation in politics, in morals, and in beliefs.

The choice of words is revealing here. Not just politics, a realm conservatives were once comfortable restricting themselves to, but "morals" and "beliefs." And not revolution or reform but "reformation." Kristol's conservatism is happy with the vocabulary of religious war. In the 1980s, the outlawing of abortion was framed in the somewhat liberal terms of saving human life, and protecting human rights. And that is why a smattering of left-leaning intellectuals also signed on as anti-abortion advocates. But in the 1990s, the conservative emphasis changed. Now the banning of abortion was linked primarily to an attack on the Supreme Court's judicial activism in other areas as well (prayer in schools, women's equality, and gay rights foremost among them) and to the more general sexual liberty of the society as a whole. Abortion became central to a reassertion of what Kristol called "traditional mores" and of "natural distinctions" between the sexes. It was not unrelated to the Lewinsky obsession and the anti-gay crusade. In fact, it was the anchor of both.

The notion of human freedom as the defining characteristic of conservatism—the characteristic championed by Barry
Goldwater, Margaret Thatcher, and Ronald Reagan—slowly
shifted toward an insistence on the "remoralization" of America,
and, indeed, the world. "Remoralizing" the world also meant
aggressively exporting American-style democracy where possible. Conservative intellectuals began to sound like the leftists of
old, bemoaning abuse of freedom, demanding that government
intervene to protect people from its baleful consequences, and
speaking constantly of ideas, ideologies, and abstractions.

Of course, because so many of the neoconservatives once
hailed from the left, this imprint was unsurprising. It mani-
fested itself in the structures of old-left activity, and still does:
the magazines and journals dedicated to the correct line; the
messianic faith in the capacity of politics to transform the world;
the infighting, and the incessant definition and redefinition of
ideology; the idolization of a supreme, largely inerrant leader,
whether he be called Reagan or Bush. One of the best descrip-
tions of this evolution actually occurred in *The Standard* in the
late 1990s, expressed by a conservative writer with a bent for
subtlety, David Brooks, now a columnist at the *New York Times*.
Brooks put his finger on a critical shift in conservatism:

> It used to be liberals who railed against the complacency of
> the American electorate, but now it's conservatives who long
> to see a little more mass outrage. It used to be liberals who
> based their politics on abstract notions more than concrete
> realities, but now it's conservatives who like to emphasize
> that ideas have consequences. It used to be liberal intellec-
> tuals who longed for the drama and turmoil that put them
> center stage, but now the habits of the New Class, both good
> and bad, have migrated rightward.

THE CONSERVATIVE SOUL · 141

He was onto something. Central to the new fundamentalist ferment was the magazine *First Things*, founded by Richard
John Neuhaus. Reading through the pages of the magazine,
more prominent in the new millennium than it was in the 1990s,
one begins to understand why conservatism as a political movement has become, in many ways, a somewhat strained version
of a religious revival. A key phalanx of its intellectual gurus has,
in fact, abandoned the secular underpinnings of the American
constitutional experiment as a whole. And, indeed, the intellectual basis on which theoconservative politics is built is a radically theocratic reinterpretation of the Constitution itself.

"The great majority of those who signed the Declaration [of
Independence]," Neuhaus wrote in the November 1997 issue,
"and of those who wrote and ratified the Constitution thought
themselves to be orthodox Christians, typically of Calvinist
leanings. It never entered their heads that in supporting this
new order they were signing on to a minimalist creed incom-
patible with their Christian profession." This is a big historical
stretch—but a necessary one to insist on closing the gap between church and state in order to save America and the world
from what Neuhaus, George, Santorum, and Pope Benedict
XVI regard as the "culture of death." While there should be
no established church, Neuhaus argued, that didn't mean that
Christianity should be forbidden from dominating the public
ethos and institutions of American government. For these theoconservatives, the whole concept of a "neutral public square" is
a delusion. Santorum called it a polytheistic anarchy; Neuhaus's
term for it is "secular monism." Such monism amounts, in Neuhaus's view, to "the exclusion of the deepest convictions of most
Americans from our politics and law" a state of affairs he described as "perverse pluralism."

What of non-Christian minorities in this divinely sanctioned order? What of those who do not, for example, adhere to Kenneth Starr's view of "private family life," or who do not even adhere to Christianity? Neuhaus insists that he doesn't want "a sacred public square but a civil public square." Nevertheless his main guarantee that non-Christian minorities would not be excluded from power is that Christians themselves would be urged to exercise "renewed opposition to every form of invidious prejudice or discrimination." The guarantee of minority religious freedom, in other words, would no longer be constitutional protection, but majority benevolence.

It is therefore unsurprising that the object of this group of thinkers deepest animosity was and is the Supreme Court, the critical enforcer of what the magazine dubbed "thinly disguised totalitarianism." The court's deference to cultural pluralism and its defense of individual conscience and a right to privacy against communal and religious authority is anathema to the theoconservatives. But even the most dogged critics of the theocons failed to anticipate the lengths to which they would go in pressing their case. In what became an infamous ideological fracas a decade ago, *First Things* actually argued for seditious activities on the part of conservative Christians "ranging from noncompliance to civil disobedience to justified revolution" against what it called, with echoes of the new left, the morally "illegitimate" American "regime." And you can see their point, from a fundamentalist perspective. The American constitution has been interpreted to permit the conscious murder of millions of innocent human beings in legal abortion. If you take the zygote as a fully formed human person, how do you not argue for taking up arms against a murderous tyranny of men and women in black robes? How, indeed, can you be a loyal citizen in a regime committed to the equivalent of another Holocaust every six years or so?

III

THE NARROW ELECTION OF GEORGE W. BUSH IN 2000—an electoral college victory that nonetheless left him with considerably fewer votes than his opponent—was the moment the new fundamentalists had been long waiting for. In the primaries, John McCain's surprising surge in New Hampshire forced Bush to an even closer alliance with the religious right than he might otherwise have preferred. Any illusions about the candidate's affiliations and loyalties were dispelled by his early appearance at Bob Jones University, a place of unrelenting Protestant fundamentalism, with a policy at the time opposed to interracial dating. It was also ferociously anti-Catholic, which gave the Bush campaign some heartburn as it tried to win the Catholic vote on socially conservative grounds as well. But by allying with the most extreme of fundamentalist elements, Bush managed to win the South Carolina primary and thereby the nomination.

Very soon the staffing of the administration was notable for its recruitment of born-again Christians. Fred Barnes, a born-again reporter, argues that the most influential members of Bush's inner circle share the president's faith. A critical member of the new administration was speech-writer Michael Gerson, a born-again fundamentalist who believed powerfully in the role of government to save souls, change the culture, and help the needy. Before long it also became clear that, like many fundamentalists, the president had some core beliefs that were impervious to any empirical qualification.

A case in point: Lower taxes were always good, regardless of the economic climate. In the campaign, the president defended his tax cuts as a way to ensure that the burgeoning surplus he inherited would not be swallowed up by the government. When the economy sank in the wake of the bursting tech-bubble and September 11, the president defended the tax cuts on opposite grounds: they were necessary as a Keynesian demand stimulus in a contracting economy. When deficits soared in his first term, Bush cited keeping the tax cuts as a way to prevent a new recession, and thereby prevent future deficits. It soon became apparent that the tax cuts were simply a matter of faith, unrelated to any empirical context or consistent rationale. Even if you supported the tax cuts, as I did and do, it was impossible logically to support every argument the president made in their defense, because they were mutually contradictory, and constantly changing. Try this experiment: Under what circumstances would President Bush actually raise taxes? Not for war; not for soaring deficits; not even with annual growth nearing 4 percent. Only one thing you could say for sure: the president never defended cutting taxes as a way to curtail the power and scope of government. Nor did he ever insist on balancing the government's books to keep the next generation out of perilous and mounting debt. That sort of conservatism was over.

The proof of that came in the staggering expansion of government power and spending under the new president. From the very beginning, the spigots of public spending were unleashed. In 2005 the federal government spent around $22,000 per household—up from a little under $19,000 in 2000. Total government spending increased by an astonishing 38 percent since 2000. This wasn't all about defense and homeland security. According to the Heritage Foundation, a conservative think tank, federal spending on housing and commerce jumped 58 percent, community and regional development 324 percent, Medicaid 49 percent, and international affairs by some 111 percent. One of George W. Bush's closest confidants and former longtime chief of staff, Andy Card, described the president's own vision of his role as president: "It struck me as I was speaking to people in Bangor, Maine, that this president sees America as we think about a ten-year-old child," he said. "I know as a parent I would sacrifice all for my children." In Bush's case, paternalism wasn't a metaphor. It was a commitment worth trillions of dollars of other people's money.

But none of this captured the full scope of the new expansion of government. Most of the president's tax cuts were backloaded, with their real costs shunted a few years into the future. The same could be said for the president's championing of a dramatic new expansion of Medicare, the government health-care program for the elderly. Put those together and peer toward the horizon and the bankrupting of the American government comes into sharper focus. According to the U.S. Government Accountability Office, in the year 2000, the United States government was committed to spending $20.4 trillion more in the future than it had any chance of taxing. By the end of one term, President Bush had more than doubled that to $43.3 trillion. That was because the welfare state was committed to providing liberal prescription drug entitlements to the largest generation in American history, the baby boomers, and that very generation was on the verge of retirement. In the late 1990s, the Clinton administration had proposed using the brief moment of government surplus to pay off the public debt and to shore up social security. Bush was having none of that. Borrowing at a furious rate, the new administration came up with the biggest new entitlement in a generation, and provided absolutely no way to finance it. No president had presided over such an increase in government spending, present and future, since Franklin Delano Roosevelt. It is safe to say that not even the most liberal Democratic president would have gotten away with it.

The scope of the government's largess was matched only by the newly emboldened defense of the morality of such expenditures. In his 2005 book, *It Takes a Family*, fundamentalist Republican Rick Santorum offers a plethora of government initiatives to remoralize Americans along lines supported by his own religious faith. He backs much tougher divorce laws for couples with children; he wants government programs for "healthy fatherhood"; he supports government programs that encompass "premarital counseling, couples counseling, conflict resolution, and parenting classes." Santorum put his name on more than 150 separate pieces of legislation in the 2004 Congress in its first year—far more than any other senator. Bush clearly agreed with Santorum's activist logic. It was compassionate to care for the elderly, and so the Medicare prescription drug law was foisted on a leery Republican caucus in Congress. It was Christian to help the needy after the devastation of Hurricane Katrina. "It's going to cost whatever it costs," the president explained. It was the right thing to spend billions on AIDS drugs for the poor in Africa, and so the money was allocated and some of it even spent.

A brief perusal of President Bush's State of the Union addresses illustrates his general sense of what government is responsible for. In 2001: "Values are important, so we have tripled funding for character education to teach our children not only reading and writing, but right from wrong." In 2003: "I propose a $450 million initiative to bring mentors to more than a million disadvantaged junior high students and children of prisoners."

In 2004: "I propose an additional $23 million for schools that want to use drug testing as a tool to save children's lives. . . . We will double federal funding for abstinence programs." The president's key new domestic initiative was to direct public funds to religious groups to perform social work, including counseling, mentoring, family support, and moral guidance. He was also deeply involved in regulating the economy—deploying selective steel tariffs, adding layers and layers of new complexity to the tax code, and increasing the budget for government regulators by a cool 46 percent in five years. Despite pledging "fiscal restraint" in his reelection campaign, the Bush administration's planned spending for 2005 was 8 percent more than in 2004, and its plans for 2006 were 9 percent more than 2005. Outstripping inflation and revenue, the Bush spending policy seemed to rest on the idea that debt was an economic irrelevance. "Deficits don't matter," was the mantra of the Bush officials. The record unequivocally shows that they meant it.

And alongside a massive expansion of government spending came the inevitable corruption. When the vast sums I've just mentioned are at stake, they act like a homing signal for every sleazeball and lobbyist in the country. The number of registered lobbyists in Washington doubled from 2000 to 2005, according to the reporter Jeffrey Birnbaum. His explanation? "In the 1990s, lobbying was largely reactive. Corporations had to fend off proposals that would have restricted them or cost them money. But with pro-business officials running the executive and legislative branches, companies are also hiring well-placed lobbyists to go on the offensive and find ways to profit from the many tax breaks, loosened regulations and other government goodies that increasingly are available."

Pork barrel spending soared, as Republican congressmen and

senators, given political permission to buy votes, vied with each other to bring federal money to their own districts. The number of local special interest projects inserted quietly into spending bills went through the roof. Here's a simple illustration: In 1998, Congress passed 1,850 such "earmarks" for transportation projects. The 2005 transportation bill contained 6,371. "The 2001-2005 period marks the transformation of the Republican Party from its traditional role as a win-or-lose guardian of limited government to that of a majority government party just as comfortable with big government as the Democrats, only with different spending priorities," says Chris DeMuth, president of the conservative American Enterprise Institute.

As with most pork barrel spending, these earmarks, moreover, often don't address real needs, but are ways to placate various local interests. And so, for example, a huge amount of federal money was siphoned to Louisiana and New Orleans in Bush's first term. But, as the *Wall Street Journal*'s John Fund reported, less than 4 percent of the total was spent on shoring up levees to prevent flooding if a hurricane hit. One other fact helps put this entire period in perspective. In 1987 President Reagan vetoed a transportation bill that included a mere 152 earmarks. In 2005, President Bush signed a transportation bill with more than 6,000 earmarks. As of this writing, President George W. Bush hasn't vetoed a single spending bill in his term of office. Given his governing philosophy, why would he?

What did this explosion of government spending have to do with conservatism? Thatcherism was renowned for its uncompromising fiscal austerity. Reagan kept domestic spending to far lower levels than George W. Bush and even raised taxes in his first term to grapple with deficits. The Gingrich Republicans, following Ronald Reagan's example, supported a balanced budget amendment to the Constitution. The first president Bush sacrificed his political future on the altar of fiscal responsibility. His son, it appears, has vowed not to make the same mistake. Part of the reason for the spending explosion was a strategic decision to woo voters previously part of the Democratic base: seniors, especially. Part was to solidify the support of religious groups. Part was the shift of Southern and especially Texan political culture to Washington—a culture steeped in personal favors, government handouts and LBJ-style palm-greasing. And part was what always happens when you don't actively try and stop government spending. It has a momentum all its own.

But none of this explains the dramatic nature of the shift, except that the new big-government conservatives really did believe in the state as a moralizing, regulating, paternalist force for improving the lives of its citizenry. This philosophy, allied with big tax cuts, was fiscally unsustainable in the long term. But it was politically manageable in the short term, which was the only horizon relevant to the new president and his advisers. They wanted to signal a new kind of conservatism, one comfortable with big government, and willing to use it to advance morality, express compassion, and reward allies and supporters. Building a new coalition—tax cuts for business and the wealthy, entitlement expansion for the middle classes and poor, public money for religious groups—was a more pressing concern than mere fiscal responsibility and limited government. Measured in terms of dollars per votes, the strategy was not successful. Despite his record of lower taxes and massive spending, the president still eked out only 51 percent of the electorate in 2004, up from 48 percent in 2000. Twenty trillion dollars of new debt bought relatively few votes in the end. The Medicare Drug Entitlement may even have ended up costing the president votes, as its complexity alienated elderly citizens.

.Perhaps what best helps explain this strategy is inherent to the fundamentalist psyche. What you see in the mind-set of President Bush is an absolute commitment to certain, often laudable, goals—helping the poor or lost, protecting the country from harm, preventing evil domestic and foreign—with a commensurate lack of interest in the means of making good on such commitments. What matters to the fundamentalist is the purity of his motives, not the messy weighing of outcomes, the adherence to cumbersome procedures, the worry about unintended consequences, the irritating follow-up of initiatives launched in a blizzard of optimism and rhetoric. It is unsurprising, then, that the faith-based charities program, Katrina reconstruction, and many of the complex tasks required to better secure the country from terrorist attacks withered on the government vine in the Bush years. Follow-up was not as important to this president as vision.

Others explain the Bush mojo in slightly different terms. "Conservatism is the right political philosophy," Bush adviser Pete Wehner told reporter Fred Barnes, "but it can be the wrong political temperament." And so conservative ends—national security, strong families, a growing economy—were promoted by radical, bold initiatives that implied a belief in the power of government to effect enormous change in the real world. In his own 2000 campaign book, Bush was clear enough about the relationship between his faith and his governing instincts. He wrote that his job as executive was "to set [the] agenda, to articulate the vision, and to lead." The notion that Bush's faith has him waiting each day for instructions from the Almighty is a silly and prejudiced one. Bush's faith has a more indirect but just as striking impact on his governing style. It empowers his boldness, permits him to focus on ends rather than means, and prioritizes vision over competence and follow-through. Above all, it allows him not to worry when his initiatives run aground, or prove counterproductive, or violate long-held ethical boundaries. He has done his job, as he sees it. He has led the way.

The fundamentalist, as we have seen, is impatient with those who claim that he must abide by certain limits in the expression or application of his religious or moral beliefs. And so the notion that compassion should or could be limited by fiscal reality has not been evident in the new Republican rhetoric. Similarly, as we shall see, the notions that the world can be redrawn by fiat, or that democracy can be created anywhere in the world, regardless of culture or tradition, were integral to the Bush doctrine, a term with literal application in this case. No one who had any sense of the limits of government, the paradoxes of power, the difficulties of easy solutions would ever say such a thing as "it's going to cost whatever it costs." The true believer, on the other hand, will tend to dismiss cramped qualifications, doubts, worries, and conditions as mere impediments to virtue, or as subtle expressions of disloyalty or, worse, doubt about the direction of policy. Such doubt has to be banished because it can cripple effective fundamentalist governance. What matters is boldness. Caution is another word for fear. The Lord will provide. Or America will somehow miraculously disprove the old economics of fiscal balance, or the hoary lessons of global history.

Domestically, two issues showed how radical the new fundamentalism really was. The first was the burgeoning debate about the role of homosexuals in society. Bush's adult years were filled with the repercussions of profound social change on the subject. Gay issues percolated through the media; AIDS forced the topic onto the table; the gays-in-the-military debate rocked the Clinton administration. Above all the issue of marriage or partnership rights for gay couples became a central front in the culture war. In 2000 Bush suggested he was a moderate on the issue, meeting with several supportive gay Republicans, and generally demurring whenever the subject came up.

In various states, sundry court rulings and voter initiatives had led to a dynamic shift in legal recognition of gay couples. Many states had banned marriages for gays; others had come up with some kind of "civil partnership" arrangement; and a single state, Massachusetts, actually made civil marriage legal for gays in 2004. There were various possible responses to this at a national level. But the first had already been accomplished. President Clinton signed the Defense of Marriage Act in 1996. That law banned any federal recognition of gay couples, and reasserted the longstanding legal fact that states had the right to ignore marriages legalized in other states.

President Bush, however, went much further. He endorsed an amendment to the federal Constitution that would make it unconstitutional for any state even to accord civil partnerships to gay couples, let alone marriage. It's worth noting here that several other conservative positions had been ruled out in advance: that marriage is a conservative institution that should include gays; that states should be allowed to figure out their own marriage policies as they had done for decades; that no action need be taken as long as the Defense of Marriage Act remained on the books; or that conservatives could support civil unions or halfway measures that could grant gays some but not all the rights of heterosexual marriage.

In many state constitutional amendments, again actively promoted by the Republican Party, gay couples were denied any benefits or protections at all. One such amendment was critical to increasing fundamentalist voter turnout in Ohio, a key swing state in the 2004 election. Another, in Virginia, even tried to ban private legal contracts between two members of the same sex who might be in a relationship. Judges—many liberal, some conservative—were described as "activists" or "extremists" if they applied their state constitution's guarantees of equal protection to gay couples. The rhetoric was, again, extraordinary. Including gays in the civil institution was described as the "abolition" of marriage, even though no one was proposing to change or amend heterosexual marriage rights one iota. Nonetheless, the Federal Marriage Amendment failed badly in the Senate, as most expected. But it signaled an uncompromising position for the new Republicanism. On matters vital to its fundamentalist base, the Republican Party would brook no compromise, accept no half-measures, and take the most drastic step imaginable: amending the very Constitution to deliver religious purity to the base. Bush formally met with no openly gay people in his term of office, said that such individuals would not share his philosophy, and publicly described civil marriage as a "sacred institution," conflating any distinction between the secular and the religious.

It was only in his second term, however, that the broader public saw the full ramifications of fundamentalist rigor in social policy. A minor case in Florida of a woman in a persistent vegetative state for fifteen years exploded onto the national consciousness. The Terri Schiavo case revealed something profound about the new conservatism. Old conservatives would have been reluctant to intervene politically in a horrifying family dispute. They would have been comfortable letting local courts or state law govern the case. And they would have acquiesced to due process, whatever qualms they might have had about the details. Today's fundamentalists, by contrast, could see little nuance in the Schiavo case: scant concern for family prerogatives, state law, judicial review, and all other painstaking proceduralism. The fundamental truth for them was that Schiavo was being murdered. A woman who had been in a persistent vegetative state for fifteen years was, for some, indistinguishable from a healthy adult. Her husband's legal rights were challenged and his allegedly sinful private life subjected to brutal scrutiny. The state legislature, governor, and then the federal Congress were cajoled to intervene. The president flew back to Washington from vacation to sign a piece of legislation designed for one specific case. The notion that government works best when it is closest to the people was abandoned. As in the marriage cases, and the medical marijuana debate, the Bush administration demonstrated that it had no interest in abiding by states' rights if those states violated the tenets of Christian fundamentalism.

Moreover the fundamentalist conservatives were able to corral the most powerful men in the Republican Party to do their will. Or rather: The most powerful men in the Republican Party were already signed-up members of the fundamentalist right. The major organs of conservative opinion—*The Weekly Standard, National Review*—both backed the fundamentalist position. *The Weekly Standard* published a long essay arguing that even if Terri Schiavo had had a living will that had cited her desire to be allowed to die if in a permanent vegetative state, she should still have been kept alive indefinitely. Morality trumped autonomy. Indeed the whole notion of individual autonomy was deemed a threat to what conservatism was seeking to defend. The judge in the Schiavo case was vilified, along with the rest of the judiciary. Then-House majority leader Tom DeLay promised retribution against the judges who ruled in the case. Senator John Cornyn made a speech, which he subsequently partly retracted, saying that decisions like the Schiavo ruling made violence against judges more understandable. A conservative conference in Washington heard participants calling Justice Anthony Kennedy's jurisprudence "satanic." James Dobson compared the Supreme Court to the Ku Klux Klan. A religious right conference baldly declared that the Democratic Party was fighting a war against all "people of faith." It was blessed by the Senate Majority leader, Bill Frist.

How were limited government conservatives supposed to respond to these fundamentalist incursions? A few dissented. The Schiavo case seemed finally to embolden traditional conservatives into defending due process and limited government. Some even defended the filibuster as an essential tool for limited and divided government. But they failed to blunt the fundamentalist position within the Republican Party, or dilute the venomous attacks on the judiciary that followed. On the marriage issue, even those with openly gay offspring, such as the vice president, were forced to toe the party line, with excruciating moments of personal discomfort in the presidential debates. The insistence by the religious right that homosexuality was a psychological disease that required treatment even forced the president to avoid ever using the words "gay" or "lesbian" or "homosexual" in his public speeches. Even to recognize the existence of gay citizens was too much for the fundamentalists. Their theological absolutism meant an end to process and a rush to results. No surprise, then, that the 2004 Republican Party platform called for constitutional amendments to ban all legal benefits and protections for gay couples everywhere in America. No surprise either that when South Dakota passed a law in 2006 outlawing abortion in all cases, including rape and incest, no leading Republican presidential candidate dissented. In a society with a large openly gay population and widespread abortion, this was not a politics of moderation. It was and is a crusade.

Crusades, however, are not means of persuasion. They are means of coercion. And so it is no accident that the new Republicans stress getting rid of obstacles to their objectives. The court system, which is designed to check executive and legislative decisions, became the first object of attack, to be checked by legislation, or permanently altered by applying a ruthless selection process to new nominees to the bench. Bare-knuckled character assassination of opponents was part of the repertoire: just look at the Swift Boat attack on John Kerry. The filibuster was attacked. The mass media was targeted, not simply to correct bad or biased reporting, but to promote points of view that were openly sectarian, even if, as in the case of Armstrong Williams, the administration had to pay for media directly.

Religious right dominance of the party machinery, in an electoral landscape remade by gerrymandering, means that few opponents of fundamentalist politics have a future in the Republican Party. It's telling that by far the biggest talents in the Republican Party will have a very hard time becoming its nominee for president. John McCain, Arnold Schwarzenegger, George Pataki, and Rudy Giuliani would all meet stiff opposition in the fundamentalist-dominated primaries. In 2000 McCain had said, "Neither party should be defined by pandering to the outer reaches of American politics and the agents of intolerance, whether they be Louis Farrakhan or Al Sharpton on the left, or Pat Robertson or Jerry Falwell on the right." By 2006 McCain had agreed to give a speech at Jerry Falwell's Liberty University. Six years spanned a vast and profound change in what the Republican Party had come to mean—and McCain had to bow to the new reality.

IV

BUT IT WAS IN FOREIGN POLICY THAT THE NEW CONSERvatism made its most spectacular impression. The defining event of the Bush presidency was the mass murder of thousands of innocent civilians only eight months after George W. Bush took office. The administration's response to that seismic event reveals both the strength and the weakness of the new politics of American fundamentalism.

The strengths first. What the American people needed after 9/11 was clarity, purpose, and a vision for a proportionate response. And this was what the new conservatism was almost designed for. Its sweeping faith in American exceptionalism, its natural, unforced patriotism, its ease with religious belief at a moment of national shock and mourning: all these were immense gifts to the country at the time. Maybe any president would have rallied to the occasion. Bush faltered at first, but then soared, with one of the most moving addresses to Congress ever given by an American president. Bush's capacity for the sweeping vision, the bold stroke, the simple, unifying message met its moment.

The new fundamentalism's ease with authority also helped. In crisis people look for strong leadership, for someone able and willing to take command. They do not want or need public displays of hand-wringing or doubt or prevarication. Bush didn't give them any. He made a couple of key decisions that still reverberate. He understood that the attack by al Qaeda and its supporters among the tyrannies of the Arab Muslim world was a declaration of war; and that the war might well at some point involve weapons of mass destruction. He also grasped that mere reaction to terror was insufficient. The West had to seize back the initiative, to put the terror networks on the defensive, to attack them before they attacked us again. He was also careful not to engage in any crude "war of civilizations" rhetoric. Despite his own Christian faith, and the fundamentalist Christian nature of his political base, he sagely eschewed any public thought of a war between Christianity and Islam.

But it was still impossible not to see, even in the beginning, the incipient dangers of a fundamentalist mind-set grappling with a huge, complex, and terrifying problem: Islamic fundamentalist terrorism. The absolutism of one almost inescapably triggered the absolutist tendencies of the other. 9/11 became, for the president, his second "born-again" moment. Just as a born-again Christian fixates upon a moment on which his entire life now pivots, the born-again presidency redefined itself entirely in terms of an absolute commitment to fighting an abstract enemy, easily conflated into a single entity, readily accessible to the fundamentalist psyche: evil.

Some of this was perfectly justified and understandable. The murder of innocents going about their daily lives is indeed evil. There is no other word for it. So too is the spiritual fanaticism of the suicide bomber. What was needed was both an ability to grasp this evil, but also a capacity for shrewd, calm, rational strategy in response, a capacity to resist the simple instinct to respond overwhelmingly in kind, or to deploy only blunt force as a weapon to defuse a metastasizing, complex political and religious cancer. It was also difficult for the new conservatives to fully grasp the fundamental nature of the enemy—religious fanaticism—because they were so motivated by religious zeal, albeit of a much milder variety, themselves.

The decision to take out the Taliban regime was, in retrospect, an easy one, although plenty of antiwar activists opposed it at the time and plenty of realists described the project as doomed. They were wrong. Afghanistan is still far from being a functional democracy, but life there is indisputably better than it was under Islamist tyranny; and the extreme danger of an unrestrained terror-state has been removed. Five years later, it is not premature to assess both the ends—ending the Taliban dictatorship—and the means—a limited, often proxy battle in remote and largely ungovernable terrain. Arguments can and should be made about the decision to delegate war-making in the critical Pakistan border areas. But few doubt that the old leadership of al Qaeda has been seriously wounded, if not fully dismantled.

The decision to follow this move by targeting the Saddam regime in Iraq was much more complicated. I don't intend to rehash here all the myriad arguments for and against the conflict. I was passionately in favor of it. In retrospect, I find the absolute certainty of my position at the time more than a little worrying. My concern about the potential danger, magnified by the continued shock and grief over 9/11, overwhelmed my skeptical instincts. I wasn't the only one in this category. Looking back, it might be helpful to see how this decision illustrates the promise and danger of political certitude; and how the new fundamentalism contributed to it, and helped compound errors that might otherwise have been avoided. In that my own errors are not easily distinguished from the Bush administration's. And that is why they bear revisiting.

The first error was an empirical one. A critical casus belli was the alleged existence of stockpiles of weapons of mass destruction and Saddam's clear intent to use them in the future, or hand them off to terrorist groups. With hindsight, that judgment was incorrect. Even though the error was committed by every major intelligence agency in the West, the Clinton and Bush administrations, and virtually every expert on the subject, it was still an unequivocal miscalculation on a huge scale. Could it have been avoided? A layperson surely cannot be blamed for taking the word of agencies whose material was, of necessity, secret. But a president, tasked with making a momentous decision, should surely have been extremely persistent in ensuring that the data were unimpeachable. The need for the appearance of this was why Secretary of State Colin Powell was selected to make the case at the United Nations. He was there to reassure the doubting that this was an empirical judgment, made by professionals, and not an ideologically driven hallucination.

But it *was* an ideologically driven hallucination. There's little doubt now that the Bush administration was anything but skeptical with respect to the data. If anything, the pressure within the administration was against those raising doubts about the evidence. The vice president's office, in particular, was furiously skeptical not of the data but of the intelligence bureaucracy's penchant for caution in risk assessment. There's a reason for that. In 1990 the first Bush administration was shocked at the extent of Saddam's WMD program. The second Bush administration, in the wake of 9/11, was terrified of making the same error. That doesn't excuse the CIA director's now-famous assessment of the existence of Saddam's WMD program as a "slam-dunk." But it helps explain the atmosphere in which such a judgment could have been made.

Misjudgments fostered in part by closed-mindedness may

be regrettable, but they are also imaginable in many administrations. The difficulty in getting international consensus for war was equally frustrating, but explicable given the enmeshment of French, Russian, and German business interests in Iraq. All of this, in other words, is understandable, and requires no theory of the new conservatism to explain it. What is not understandable, however, is the almost nonexistent preparation for war made by the Bush administration, the stunning lack of foresight about the dangers of Iraq after invasion, and the continued reluctance of the administration to adjust once clear mistakes had been exposed. Given the stakes involved and the immense difficulty of the task, it is still difficult to explain a war policy of what can only be called intransigent recklessness.

Some of the errors can be attributed to the fog of war, to the inevitable mismatch between theory and practice, between war plans and an actual conflict, taking place in a deeply divided country sealed off for years from most outside contact, and exhibiting what can only be called post-totalitarian syndrome. No one should expect perfection. But what we witnessed was something far more disturbing: a refusal to account for reality, to acknowledge error, to prepare for all contingencies. In searching for an explanation for that, we have to return, I think, to the kind of conservatism George W. Bush had internalized.

In that worldview, what mattered was the ideological analysis: good versus evil. What mattered was the assertion of the United States' right to act alone if necessary to defend its own security. What mattered was the zero-sum analysis that we had to choose between war against Saddam and a potential mushroom cloud in an American city. It was this rigid and abstract analysis that essentially abolished the idea that the war was subject to rational debate. A columnist asked the president some time before the war if he was serious about invasion. Bush responded: "I don't think we have a choice, do we?" In an interview after the war's inception, the president was asked if he viewed the war in retrospect as a war of choice or a war of necessity. The president looked as if it had been the first time he had even asked himself that question. The fundamentalist makes his mind up instantly, makes the fundamental decision, and cannot, by necessity, stop short at a later date and ask himself if he's right. Such second-guessing undermines his entire worldview. It threatens his inner psychological core.

And this narrative—amazingly—continued throughout the postinvasion anarchy. Glaring and obvious problems were responded to by a reiteration of the theory for the invasion. Democracy is on the march. Freedom is at hand. The insurgency is in its "last throes." The notion that Saddam and the Sunni minority might respond to a conventional invasion by launching a guerrilla war had not, apparently, occurred to the White House. There is no more compelling metaphor for the mismatch between ideological certitude and complex reality than the president's victory speech under the banner "Mission Accomplished." We now know that those on the ground, like Paul Bremer, immediately saw that there were far too few troops to maintain order or provide stability for any democratic transition. But Bremer—who wanted to triple the forces on the ground—was ignored. Such concerns voiced before the war by General Shinseki were dismissed out of hand. In the postinvasion period, any criticism of the operation was equally described by many Bush supporters as a form of treachery or alliance with the enemy, rather than a simple, empirical question about what we were doing and whether it was effective.

In the wake of growing chaos, murder, and political drift, the

Bush presidency merely insisted that nothing was wrong. All questions about troop levels were deflected to the commanders in the field who reported to a president who didn't want to hear about difficulty. According to Fred Barnes, the weekly conference calls with the generals were not exactly opportunities for differences to be raised or questions asked. "Bush liked crisp sessions without whining or complaints," Barnes wrote. "Once he had to interrupt a discussion of troop rotation to say, 'Stop the hand-wringing!'" In the presidential debates, Bush was asked whether he could name a single mistake in his presidency thus far. He wouldn't. Or perhaps it is more accurate and unnerving to say: he couldn't.

Part of this brittleness can be understood as public relations. War leaders do not want to be seen second-guessing strategy in public. Much of the opposition in America would have jumped on any concession to reality by the president and used it against him. But again this doesn't fully explain the rigidity of the Bush White House, its imperviousness to empirical criticism, its insistence on the inerrancy of its leader, and its ruthlessness toward critics. What does help explain it is the fundamentalist mind-set. A strong inerrant leader is typical of such religious groupings; deference is regarded as the natural response to such a hierarchy; criticism is immediately conflated with sin or weakness or treachery. Loyalty, however, is always valued—even when it appears ludicrous. One of the more bizarre moments in a presidency firmly detached from empirical reality was the decision to give the highest civilian honors possible to the CIA director who had made the biggest intelligence misjudgment in modern history and to a de facto consul of Iraq, when it was reeling from the consequences of American short-sightedness. The most unsettling aspect of the spectacle of the Presidential Medals of Freedom being given to George Tenet and Paul Bremer was that the president showed absolutely no sign of appreciating even a scintilla of the irony.

Bush's followers often spoke of him in terms usually reserved for spiritual leaders—even during the chaotic months after the Iraq invasion. One supporter, John Hinderaker, whose blog Power Line remains a favorite of the Republican right, wrote the following words in July 2005: "It must be very strange to be President Bush. A man of extraordinary vision and brilliance approaching to genius, he can't get anyone to notice. He is like a great painter or musician who is ahead of his time, and who unveils one masterpiece after another to a reception that, when not bored, is hostile." By that point, close to thirty thousand Iraqi civilians had been murdered while the United States was responsible for Iraq's security.

"Stuff happens" was Donald Rumsfeld's famous response to what Paul Bremer called the "industrial-strength looting" that occurred after the fall of Baghdad. Among the sites looted: weapons sites that the administration had, before the war, cited as one of the rationales for the entire enterprise. None of this could be acknowledged, because, in the words of Bush-loyalist Fred Barnes, it would "undermine the mystique of the presidency and erode the president's effectiveness as the nation's leader." Notice the descriptive term: *mystique*. It is usually deployed to describe popes and monarchs, not presidents in a republic.

And with the mystique of the inerrant presidency came a new theory of his constitutional powers. It became clearer and clearer throughout the Bush presidency that the White House backed the idea of the executive as a war-making authority that could be checked by no relevant law and no relevant treaty. The presidency, according to this view, operated in a realm outside the law when it was tasked with defending the nation from a threat to its security. And so the president was fully within his rights to ignore, say, the 1978 law that mandated that any wiretaps of domestic citizens be run through a secret court in order to get a warrant. Some argued that the law needed to be amended to take into account the revolution in information technology that had occurred in the last three decades. But the administration rebuffed congressional attempts to revise the law—because, in its view, the president had the right to ignore the law in the first place.

The president also put in place a novel way to bypass laws with which he disagreed in the area of national security. He dramatically altered the already novel practice of presidents' attaching "signing statements" to laws upon their passage. A few dozen such statements had been issued in the two centuries before Ronald Reagan's presidency. Reagan attached seventy-one in eight years, explaining why he hadn't vetoed a bill or providing boilerplate to claim credit for it. Bill Clinton appended around one hundred in eight years. In five years, President Bush qualified over five hundred legal provisions, almost all of which affected the executive branch's management of its own affairs. He reserved the right to ignore plain congressional statutes if he believed they might impede his authority to run a war as he saw fit.

Republican senator Lindsey Graham was insightful enough to see the full import of this. "If you take this to its logical conclusion, because during war the commander in chief has an obligation to protect us, any statute on the books could be summarily waived," Graham argued. It is very hard to disagree with him. But when a presidency is understood in semi-religious fashion, when his authority as commander in chief cannot be abridged by any check or qualification, such signing statements are perfectly logical. If the threat is all-consuming, and the president is solely tasked to defend a nation of helpless children from evil, then of course Congress cannot impede his powers. Too much is at stake. The only check on the president's power was an election. Hence Bush's stunning description of the 2004 election as his accountability "moment." Once that moment had passed, the president was accountable only to himself. This situation, moreover, was not merely an emergency expedient, designed to expire in a short period of time. Since the war was defined as unending, the new power of the presidency promised to alter the very dynamics of the American republic toward a pseudo-monarchy.

Perhaps this is better understood by comparing different possible approaches to a democracy in wartime. Churchill made a point of constantly reminding the British of their vulnerability, and their failures. He insisted that everyone make real sacrifices in the war effort. He brought opposition leaders into his war coalition. And he regularly assessed the state of play in detail in Parliament. What makes democracies good at warfare, after all, is their ability to question strategy and tactics, and to alter war methods to cope with new contingencies or unforeseen events. Churchill saw the British people—of all political persuasions— as key conscripts in a war. Bush, in contrast, asked of Americans merely that they go back to shopping, remain vigilant, and vote for him to protect them. He ignored the Congress as much as possible, spent critical months in the Iraq war touring the country backing social security reform, and addressed any and all questions about the conduct of the war by reiterating merely that it was "hard work," and that all key decisions were made by the military and not him. Again: what you see here is a fundamentalist psyche at work. Only one patriarchal leader can be trusted; and he is inerrant. The role of followers is not to question but to submit. And in a Manichean war against evil, any means were justified—however dubious their legality, however dangerous the precedent.

The most striking aspect of this new executive power—free from all legislative, judicial, or international checks—was the Bush administration's secret but undeniable endorsement of torture by American soldiers and CIA personnel. It would be hard to overstate the radicalism of what was done. But the Bush administration began the war with a specific decision not to abide by the Geneva Conventions if "military necessity" made them, in the view of the president, dangerous to national security. Given this new flexibility, the CIA set up a network of secret prison sites for the interrogation of military detainees, and finessed techniques that clearly contravened U.S. law and the Geneva Conventions. The use of forced nakedness of detainees, of "stress positions" that made it difficult for prisoners to breathe or stand, of controlled hypothermia and extreme heat to break down prisoners, of unmuzzled dogs, of sexual abuse and even rape, and repeated beatings: all became widespread in the war on terror.

The incidents of abuse were recorded everywhere, according to the government's own reports: from Guantánamo Bay to Afghanistan, Baghdad, Basra, Ramadi, and Tikrit and, for all we know, in any number of hidden jails affecting "ghost detainees" kept from the purview of the Red Cross. They were committed by the marines, the army, the military police, Navy SEALs, reservists, special forces, and on and on. The use of hoods was ubiquitous; there are even a few examples of electric shocks. Many of the abuses seem specifically tailored to humiliate Arabs and Muslims, where horror at being exposed in public is a deep cultural artifact. Whether random bad apples had picked up these techniques from hearsay or whether these practices represented methods authorized by commanders grappling with ambiguous directions from Washington is hard to pin down from the official reports. But it is surely significant that very few abuses occurred in what the Red Cross calls "regular internment facilities."

Some soldiers objected. One West Point graduate testified that he had seen routine beatings of detainees, often for sport. Others spoke of gruesome practices: of the beating to death of an innocent man at a place called the "salt pit" in Afghanistan; of immersing detainees in freezing water until they confessed some kind of crime or terror connection; military e-mails confirm that General Geoffrey Miller, who had pioneered "coercive interrogation techniques" at Guantánamo Bay had been sent to Abu Ghraib to "Gitmoize" it. Among the abuses cited by the Pentagon's own inquiry is the following:

On another occasion DETAINEE-07 was forced to lie down while M.P.'s jumped onto his back and legs. He was beaten with a broom and a chemical light was broken and poured over his body. . . . During this abuse a police stick was used to sodomize DETAINEE-07 and two female M.P.'s were hitting him, throwing a ball at his penis, and taking photographs.

Here's the testimony of an army interrogator in Afghanistan:

When the Navy SEALs would interrogate people, they were using ice water to lower the body temperature of the prisoner and they would take his rectal temperature in order to make sure that he didn't die. I didn't see this, but that's what many, many prisoners told me who came out of the SEAL Compound, and I also heard that from a guard who was working in our detention facility, who was present during an interrogation that the SEAL had done.

The interrogator asserted that he too had used hypothermia, although not as clinically as the Navy SEALs. Incidents of this kind occurred throughout the theater of war and, for all we know, are continuing. The *Washington Post* reported, based on leaks from within the CIA, that "waterboarding" had been officially authorized in at least eleven cases of al Qaeda detainees. Here's the CIA's definition of waterboarding: "The prisoner is bound to an inclined board, feet raised and head slightly below the feet. Cellophane is wrapped over the prisoner's face and water is poured over him. Unavoidably, the gag reflex kicks in and a terrifying fear of drowning leads to almost instant pleas to bring the treatment to a halt." Even the most passionate of the administration's defenders cannot argue that this is not "cruel, degrading and inhuman" treatment, as banned under the Geneva Conventions and American law. What they did assert was that waterboarding wasn't "torture," as they defined it.

The definition of "torture" in the U.N. Convention to which the United States is a signatory is the following: "any act by which severe pain or suffering, whether physical or mental, is intentionally inflicted on a person for such purposes as obtaining from him or a third person information or a confession . . . when such pain or suffering is inflicted at the instigation of or with the consent or acquiescence of a public official or other person acting in an official capacity." The notion that "waterboarding" is not torture under the plain meaning of the word as well as its legal meaning is preposterous.

And yet the president continued to insist that "we do not torture." The photographs from Abu Ghraib refuted him, but he again insisted, against the overwhelming evidence, that there was no connection between the looser standards of interrogation he had endorsed and what occurred in Abu Ghraib. Again we see the fundamentalist psyche: what matters is his intent, not the empirical analysis. The president insisted that what happened at Abu Ghraib did not represent America—but the memos and documents from his own government showed that they were the logical consequence of his own decision to end decades of humane warfare in the United States military. The legal adviser who helped lower the interrogation standards, John Yoo, was more candid. In a debate at Notre Dame University with human rights lawyer Doug Cassel the following exchange occurred:

CASSEL: If the president deems that he's got to torture somebody, including by crushing the testicles of the person's
  child, there is no law that can stop him?
YOO: No treaty.
CASSEL: Also no law by Congress—that is what you wrote in
  the August 2002 memo. . . .
YOO: I think it depends on why the president thinks he needs
  to do that. . . .

Yoo works at one of the most prestigious think tanks in the United States: the American Enterprise Institute. He is absolutely sincere in believing that the executive branch can override any domestic law, any international treaty, and any moral boundary if necessary to protect national security. In a war on terror that stretches decades into the future, the new conservatism allows for a president with no checks at all on his own power as commander in chief. What might have once been a theoretical debate became a pressing reality. And within weeks of this new legal doctrine being expressed, military detainees under the control of American forces were being tortured—consciously, with premeditation, with legal cover provided. America went from being a constitutional republic, under the law, to an imperium of one man, answerable only to an election every four years, empowered to break any law and violate any moral law if he believes it is necessary for national security. If conservatism had begun as a political philosophy designed to check power, to ensure individual liberty, to protect individuals from lawless government authority, it ended in a dark room, with a defenseless detainee strapped to a board, terrified beyond most of our imagining.

CHAPTER 5

The Conservatism of Doubt

"Some impose upon the world beliefs they do not hold; others,
more in number, impose beliefs upon themselves, not being able
to penetrate into what it really is to believe."
—MICHEL DE MONTAIGNE, "Apology for Raymond Sebond"

I

THE DEFINING CHARACTERISTIC OF THE CONSERVATIVE IS
that he knows what he doesn't know.

Like the mark of a fundamentalist, this brief statement carries a world of weight and begs a lot of questions. But it is, I
believe, indispensable to the conservative disposition, the key to
understanding what conservatism can mean in modernity, the
central argument that offers an alternative to the fundamentalist temptation.

The first thing to say is that this philosophy is not warmed-
over relativism or nihilism. While the fundamentalist knows
the truth, the nihilist believes it is an illusion, that nothing is
true, and everything is valid. The conservative differs from both.
While not denying that the truth exists, the conservative is content to say merely that his grasp on it is always provisional. He
may be wrong. He begins with the assumption that the human
mind is fallible, that it can delude itself, make mistakes, or see
only so far ahead. And this, the conservative avers, is what it
means to be human.

No person who ever lived had the perspective of eternity or the gift of omniscience. Yes, Christians may want to say that of Jesus. But even the Gospels tell us that Jesus doubted on the cross, asking why his own father seemed to have abandoned him. The mystery that Christians are asked to embrace is not that Jesus was God, but that he was God-made-man, which is to say prone to the feelings and doubts and joys and agonies of being human. Jesus himself seemed to make a point of this. He taught in parables rather than in abstract theories. He told stories. He had friends. He got to places late; he misread the actions of others; he wept; he felt disappointment; he asked as many questions as he gave answers; and he was often silent in self-doubt, or elusive, or afraid. God-As-Omniscience, by definition, could do and be none of these things. Hence the sacrifice entailed in God becoming man. So at the core of the very Gospels on which fundamentalists rely for their passionate certainty is a definition of humanness that is marked by imperfection and uncertainty.

This is where a conservative starts. If we are all humans, then we each have a beginning and an end; and each of us has a different beginning and end. We see the world from where we are, and our understanding of the universe is intrinsically rooted in a time and place. We can do all we can to increase our knowledge and gain deeper and deeper insight. We can read history and philosophy; we can travel; we can ask questions of young and old; we can debate; we can pray; we can grow through the pain and amusement of daily life. But we will never fully or completely transcend where we are. And even if we could, such transcendence would render us unintelligible to those still earthbound.

There are so many stories in human history that relate this

truth about humanness, and so many parables and analogies. There is Plato's famous cave. There is the fable in Genesis of the Fall. There is Thomas More's *Utopia*. There is every play Shakespeare ever wrote and every essay Montaigne revised. All tell us that tragedy or comedy are intrinsic to the human story. God, after all, can be neither comic nor tragic. He is beyond such categories. As humans we can merely sense the existence of a higher truth, a greater coherence than ourselves, but we cannot see it face to face. We cannot fully capture it and remain human. And yet we cannot fully escape its lure either. This is either funny or sad, and humans stagger from one option to the other. Neither beasts nor angels, we live in twilight, and we are unsure whether it is a prelude to morning or night.

The mark of the conservative temperament is an acknowledgment of this twilight. Acknowledgment is the right word, I think, because it connotes neither capitulation nor rebellion. The conservative does not deny the truth of which the fundamentalist speaks so clearly. But he cannot affirm it with the same surety. He will ask: how can we know for sure? When so many proffer so many contradictory truths, how can we know for certain which one is real? Eternal truth—the ultimate form, the moment when all doubt ceases, the salvific instant when the clouded glass of human consciousness is wiped clean into transparency—is definitionally beyond us. It is what we die for. And in dying, we cease to be a human among humans.

This deepest truth about human life is therefore also a political truth, because humans live among one another, and the way we arrange that association is what politics is. The best depiction of the full implications of this is still, in my view, the first. It is in Plato's *Republic,* in the famous cave story I just referred to. In the dialogue, Socrates challenges his interlocutors, chiefly young Glaucon, a precocious young man with political aspirations, to analogize the process of searching for the truth.

Socrates describes human existence as being in "an underground cave-like dwelling with its entrance, a long one, open to the light across the whole width of the cave." But most people are in shackles in the heart of the cave, and they cannot turn their heads around to look directly toward the light, or get up and move toward it. So they see things by their shadows, and the shadows of other things and people, reflected on the wall of the cave in front of them, with light coming from behind them. In fact, this is how they see themselves. They develop a language and a discourse that revolves around this world. They name the things they see as shadows, see connections between them, interpret the sounds they make. They construct an entire universe from their limited perspective.

This, Socrates implies, is what human life is like. The cave is the world not as it is in full, but as we see it, with our limited perspective. Now, Socrates says, imagine one of these prisoners being able to get up, turn around, and look toward the light. Imagine even further that he can make it to the opening of the cave, and see reality: an entire world beyond the cave, the truth about things, illuminated not by fire and reflections, but by the sun itself.

The analogy works on so many levels. And, like all great stories, it has spawned numerous interpretations. Here's the one that makes most sense to me. Plato is telling us that seeing the truth is not completely beyond us. A few can wrestle—and have wrestled—themselves free of the bondage of illogic, prejudice, sentiment, bias, self-delusion, fear, self-interest, passion, and misunderstanding that human thought is heir to. But this is rare. A Socrates or Jesus or Muhammad or Einstein does not come every day. And even these few must necessarily find the process of arriving at the truth a difficult one. It is something, Socrates says, that "by nature" "happens to them." For the saints, it is a moment of revelation from above. In Socrates' world, it is an achievement of a life of rational effort, driven by a human longing for completeness. The nature that propels even the greatest saints and philosophers to the truth is in all of us, to varying degrees. Few humans have no interest in understanding the shadows on the wall. We all want to grasp the world we live in. Like plants, we lean toward light and away from darkness.

The genius of Plato, however, is in seeing how this interacts with human psychology. Think of yourself as one of those able to get out of your chains and stagger toward the light. Imagine that you have lived your entire life in the cave. What would it feel like to see the truth, to enter a light you had only previously grasped indirectly? It would be blinding, at first, disorienting. You would shield your eyes from the sun, squint through them, peer into places that you had previously been unable to imagine. It would upset every preconceived idea you ever had.

Its beauty and goodness would overwhelm and entrance you. That thread of reason and self-consciousness that makes man separate from animals would be flooded by truth, like a brain overwhelmed by dopamine. It would be both all encompassing and beyond description. Notice also how the ultimate truth is, perhaps, even then impossible. If you look directly at the sun—and who, in such a circumstance would want anything else?—you will be blinded. So you tend to look at the earth and see things in a new light. Or you're grateful for the night, when your vision, trained by years in twilight, is better suited for understanding what you see.

But you are pretty much alone up here. Or your company

is sparse. And you are still human. You need family and friends and community, even though the truth now calls you with an insistent voice. You may want to bring your friends and family into this light as well. At the same time, the empowering, beautiful experience of seeing the world as it is, is a terrible thing to let go. And so going back into the cave is a difficult, a painful thing. And when you get there, what will be the reaction of those in the dark?

At first, of course, you will be blinded again, your eyes used to sunlight, now having to get accustomed to darkness again. You may look odd as you come before your old friends, still blinking and breathless. And you explain all these things. You put color into the shadows, you tell them stories of wondrous things, of strange truths, of a world that makes theirs seem damp and dark and colorless in comparison.

How would they regard you? Almost all of them, after all, have made the cave their home, as it has been the home for their fathers and mothers. They have become used to it; and they understand it. And they are not wrong about it, as far as they can see. Some may be intrigued by your story. A few may even treat you like a God. But many will also think you a lunatic. It takes time for your eyesight to adjust again, and for a while you make elementary mistakes in seeing your old world as others see it, and they mock you. If he cannot see the cave correctly, why should we trust what he says about what is outside? There are, moreover, many men and women in the cave who have developed the keenest eyesight and intelligence for understanding their own twilight world. You threaten them and their status. If you are right, then they are frauds, or merely far inferior to you. And so Socrates hears a very human response to the man who returns with the truth:

Wouldn't he be the source of laughter, and wouldn't it be said of him that he went up and came back with his eyes corrupted, and that it's not even worth trying to go up? And if they were somehow able to get their hands on and kill the man who attempts to release and lead up, wouldn't they kill him?

And so Socrates predicts his own death; and that of Jesus, for good measure. Some see in Plato's dialogue a call for philosophers to rule the earth according to truth: the fundamentalist vision writ large. I disagree. Plato knew Socrates' fate. He knew what could happen when real philosophers speak the truth in a polity. And when you read his account of the perfect city "in speech," you realize, as you do in More's *Utopia* (which means "nowhere"), that the point of the dialogue is to elucidate why such a city is impossible in reality. The parable, like so much of Plato, is ironic.

The very form of writing he uses is instructive in this regard. It's a dialogue, not a proof. It's a drama, not a program. It is full of arguments that sometimes go somewhere, sometimes nowhere. Its various interlocutors have their own perspectives which the reader has to take into account in order to understand the issues involved. And every time I have read it, I have seen something new in it, an argument I found persuasive last time that didn't make much sense this time, an implication not fully pursued, a metaphor that yields new wealth.

We can speak easily and glibly of what is true, but this is what human life is actually like. A conversation about the meaning of justice leaves us with as many new questions as provisional answers; and it's necessarily conducted by human beings with different personalities, interests, ages, virtues, and vices. This is true even of a conversation with Socrates! How much cloudier and digressive must our own conversations be? Or even our own internal discussions, those moments when we confront our own assumptions as we drift into sleep, or are moved by a poem, or challenged by a new fact?

What I'm left with at the end of the story is a sense that neither the limited horizon of the cave dwellers nor the wonderment of those who have found the truth contain the key to the human predicament. The key is with Plato: because he sees both. He sees the fundamental incompatibility of certainty with humanity; of philosophy with politics; of ideas with practice. Somehow we have to live in the interstices. We can neither deny the call of the light nor the necessity of living collectively in shadows. We cannot live according to perfect ideas of the good or the true or the beautiful; but neither can we live fully as humans without engaging them.

The key, therefore, to politics and to life in general is a recognition of imperfection. The core truth of this imperfection is that what we can know to be true in our minds or souls, in the realm of ideas or faith, can never be easily replicated in the real, material, physical world, either individually or in association with others. There is, in short, a deep disjunction between perfect theory and imperfect practice. And the greatest human error is to believe that the two can be reconciled, that heaven can be reproduced on earth, that there is a truth "up there" that must or can be transposed "down here"; that the perfect city in speech can be a city in reality; that somewhere on earth, the ultimate human problems can be solved forever. What Plato is telling us is that this is an illusion. The city and the philosopher can never be joined. The pursuit of truth cannot be fully reconciled with the government of mortals, who are wedded by custom and feeling and habit and history to various untruths, myths, prejudices, half-truths, and feelings.

We are, in short, fallible creatures. We get things wrong. We differ with one another not only reasonably but unreasonably, when passion or pride or interest gets in the way. And something is always in the way.

II

TO MY MIND, THE GREATEST ELABORATOR OF THIS VIEW of flawed humanity is Michel de Montaigne, the sixteenth-century French essayist, whose distinctive voice still resonates through the centuries with a quirky yet serene tone. Like us, Montaigne lived in an age of passionate religious certainty—a certainty that provoked Catholics and Protestants to burn and murder one another, that brought new regimes to power and destroyed old ones, that wrought havoc on social order and civility, and all in the pursuit of a perfect justice and godly order. Montaigne saw all this and . . . sighed.

"Presumption," he wrote, "is our natural and original malady. The most vulnerable and frail of creatures is man, and at the same time the most arrogant. He feels and sees himself lodged here, amid the mire and dung of the world, nailed and riveted to the worst, the deadest, and the most stagnant part of the universe, on the lowest story of the house and the farthest from the vault of heaven . . . and in his imagination he goes planting above to the circle of the moon, and bringing the sky down beneath his feet."

Who on earth are we, Montaigne asks, to speak so certainly of things that, by definition, we cannot know for sure?

He looks above and below for evidence of our claims to truth. Above Montaigne recognized a universe so vast and incomprehensible that the whole idea of grasping it in a human mind seemed preposterous to him, ludicrous, comic. Each of us is but one fleeting member of one small species on a relative speck of dust, orbiting a third-rate star in a galaxy among millions of galaxies, constantly expanding into new and vast unknowns. And yet mullahs and priests and rabbis and social scientists and philosophers are claiming to have discovered the meaning of everything so perfectly that, in some cases, they even aspire to infallibility. Montaigne's response to this is simply to laugh at it. It is funny. In fact, the only thing more hilarious than the notion that any human has figured it all out is the idea that he should then proceed to tell others how to live their lives according to his divine or perfect vision.

Montaigne argues that we do not even understand the minutest of our own daily interactions, let alone the eternal verities. We are almost as blind with respect to the world below us as to the worlds above us. "When I play with my cat," he asks, "who knows if I am not a pastime to her more than she is to me?" He wonders whether other animal species don't really own the planet and understand it better, while we pass self-centeredly through. There is a mystery in everything outside ourselves, he argues, that we can never fully grasp. If this is true of a cat, how much truer of other human beings?

We judge others often simply because we do not understand, or because what we see is something we are not entirely used to, or because we naively assume that they are just like us. Some of this, scientists have begun to unravel, is hardwired into our genes; some of it is inculcated by upbringing or habit; some is a function of sudden passion or anger or attraction; some of it is a consequence of fallible senses and illogic. And no one is immune from these self-deceptions: "The same reason that makes us bicker with our neighbor creates a war between princes; the same reason that makes us whip a lackey, when it happens in a king makes him ruin a province. Their will is as frivolous as ours, but their power is greater. Like appetites move a mite and an elephant."

All we really know, Montaigne would argue, is ourselves. Even then, we are capable of self-delusion, of change through time, of shifting judgments and moods, of views that we held once that we have abandoned today, or opinions that we still hold but more weakly, more provisionally. We are internally mysterious even to ourselves. The man I am today does indeed bear striking resemblances to the boy I once was, to the moody and troubled teenager I became, to the young man who once knew things for certain, and judged things quickly and surely as they came along. Even my faith can be said to have remained relatively stable for four decades: with only one relatively brief period when I stopped attending mass, I have remained a Catholic.

At the same time, my former self exists today only as filtered through memory and time. I see home movies of myself at a younger age and feel a shock of strangeness. I read sentences I once wrote and recognize in them a different sensibility than the one I seem to hold now. I can watch a good movie at various times in my life, and each time, it seems slightly altered and new to me. Usually the differences are nuanced. But sometimes it is as if the film is completely fresh, as if I never previously understood it, as if it took me these extra years of experience and growth or even decline to see it as it really is.

Great art impacts me differently at different moments in my life. I acted in Shakespeare's *Hamlet* in high school in the part

of Osric, the foppish courtier who has one comic scene near the
end. I rehearsed just that part in isolation until the final week
of rehearsal, and never read the entire play in advance. For all I
knew, the play was a comedy, and my job was to get a few laughs.
And so I discovered the plot line as it happened in front of me
at the age of fourteen, in dress rehearsal. I can still remember
the shock as I stood onstage and watched the final scene, and
saw one character after another die in front of me. By the last
line, I was relieved to be alive, and in some shock at the sheer
bloodiness of the entire thing. And once I knew the brutal ending of the play, it was never the same for me again. I could never
be surprised by Shakespeare's plot again, and so there was one
time and one time only that I could really see Hamlet as a new
viewer was supposed to see it; and be surprised and enthralled
by its unfolding.

In college I saw a production of it where, for the first time,
I saw the full psychosexual dimensions of the closet scene between Hamlet and his mother, and the full import of Hamlet's strained, bizarre love for Ophelia. In high school, I had
not known love and barely knew sex and so I had not understood the play. Now I did—but differently, and more deeply.
In graduate school, I actually played the part of Hamlet in an
experimental production and discovered that much of what I
had previously understood about the character receded in front
of me. I found that when I tried to become Hamlet instead of
watching him, I couldn't grasp him as a human being at all. I
came to the view that Hamlet was not a person, but a kind of
physical representation of pure thought. And then, in subsequent years, I saw the play again and found it far funnier than I
recalled, and watched others play the parts in ways I had never
really anticipated. Single lines whose meaning had once seemed

obvious to me took on new shape and dimension by the mere inflection of a voice, or the pausing over a comma, or the removal of a piece of dialogue.

My conclusion? Over four decades, I changed, and the play changed. I experienced the play as a bit part, as an audience member, as Hamlet, and as an actor behind stage hearing others' scenes echo through the theater. And each time *Hamlet* was different and I was different. What makes live theater so unique—what makes it theater—is that a play and a performance and an audience exist only once in the same combination. Change anything—actor, director, audience, mood, environment—and the entire thing changes. And each experience is a kind of mystery, and the mystery includes ourselves.

If this is true of a play, how much truer is it of life itself? How much of the outside world can we truly know? We all know of our own failings in judging others' motives and characters. We get people wrong all the time. We elect a president or prime minister with our hopes up, only to discover that he isn't what we thought. We rely on a coworker for a vital task, only to discover that he isn't up to the job, or had different skills than we envisaged. And these are merely the misunderstandings of people somewhat distanced from us. We get the familiar wrong as well. Our children grow up and disappoint or astound us; our parents, people on whom we once relied for everything, turn out to be human too, and flawed.

We fall in love and experience the most intense communion with another human being we ever thought imaginable. For a while, we are transported into ecstasy, as the loneliness of being is dispelled by the presence of someone else who shares life with us, who can even create new life with us through the most intense communion of sex. And then we slowly seem to grow apart and the excitement abates, and what we thought we knew for sure turns out to be not so sure. And the disappointments turn into resentments and the intense proximity reverses itself into cold detachment. Divorce is, in many ways, simply an emblem of the limits of human understanding, or a tattered remnant of a previous hope triumphing over reason, our passions pushing away the doubts that nibble at the edges of our consciousness until they change it altogether. Or sometimes divorce isn't a failure of understanding at all. Sometimes it's just a reflection of the fact that, as time goes by, each half of a couple changes and grows in a different direction, and the people who first fell in love with one another no longer fully exist as they once were, and so the relationship falls apart, or calcifies, or becomes something else again. And this, remember, is true of someone we have focused on with all our attention, someone we have been more intimate with than anyone, someone we should have a chance of understanding and knowing fully, if we are to have a chance of understanding and knowing anyone at all.

It turns out we don't know and didn't know and cannot know for sure. Even when we do, when some other human being retains our trust and love and friendship for a long period of time, we may still not fully know her as she knows herself. If we remain partly mysterious to ourselves, there will always be some gap between ourselves and others, some lacuna to be filled, some aspect we haven't quite figured out.

All of this should make us profoundly reluctant to speak certainly of that which we cannot fully know. And there is no subject more beyond our capacity than the ultimate questions: What lies behind all this? What does God want of us? How are we to live our lives? "Our powers," Montaigne writes, "are so far from conceiving the sublimity of God, that of the works of our creator that bear his stamp most clearly, and are most his, those we understand least." Each attempt to capture a truth subsequently evades us, as a further question beckons. Each attempt to satisfy a desire subsequently slips out of our reach, as a new need instantly springs up to replace the old one. There is no stopping place here, no repose, no moment of calm. It is all in flux and opaque and evanescent. In the words of T. S. Eliot,

> *And what you thought you came for*
> *Is only a shell, a husk of meaning*
> *From which the purpose breaks only when it is fulfilled*
> *If at all.*

Those who claim to have captured the divine in a chapel or a mosque or a book have merely shown that they have not captured it. "In Socrates' opinion, and mine too," Montaigne writes, "the wisest way to judge heaven is not to judge it at all." But many look at nature and project onto it their own feelings and desires, their own beliefs about what is good and true, irrespective of the complexity and diversity of creation as it actually is. What else, ultimately, is "natural law"? You can, in other words, easily detect those who are the most lost in a fog because they are the only ones insisting they can see clearly. Montaigne elaborates:

> We cannot worthily conceive the grandeur of those sublime and divine promises, if we can conceive them at all; to imagine them worthily, we must imagine them unimaginable, ineffable, and incomprehensible, and completely different from those of our miserable experience. "Eye cannot see," says Saint Paul, "neither can it have entered into the heart of man, the happiness which God hath prepared for them that love him."

And so Montaigne turns Christianity against the fundamentalists. Their certainty, he argues, is the real blasphemy; their desire to control the lives of others the real heresy; their simple depiction of the Godhead proof positive that they do not really understand him: "When we say that the infinity of the centuries both past and to come is to God but an instant, that his goodness, wisdom, power, are the same thing as his essence—our tongues say it, but our intelligence does not apprehend it. And yet our overweening arrogance would pass the deity through our sieve." Think of the sieve as "biblical law" or "dispensationalism" or "Wahhabism." And before you start arguing, have a good laugh at the insane arrogance of it all.

In this, of course, Montaigne was not alone. Perhaps the greatest political philosopher of modern times, Thomas Hobbes, also grasped this essential truth about man's relationship with the divine:

> For the nature of God is incomprehensible; that is to say, we understand nothing of *what he is*, but only *that he is;* and therefore the attributes we give him are not to tell one another what he is, nor to signify our opinion of his nature, but our desire to honor him with such names as we conceive most honourable amongst ourselves.

It's all projection; and there is nothing wrong with that—so long as we do not mistake it for something else, namely real knowledge of something that by definition is ineffable. Plato says the same thing in the *Republic*: "That which is wholly a being is wholly knowable, while that which is in no way a being is in every way unknowable." God is beyond being. God is beyond our mortal human categories, beyond the deadliness of doing, beyond the constant failure of human achievement, which leads to yet more achievement subsequently revealed in a changing world as yet more failure. We cannot make sense of God; we can merely revere him. That last pronoun, projecting onto God a gender like humans, is itself an indication of the problem. Such are the necessary limits of our intelligence, of our very language in trying to capture something beyond language.

Montaigne also invokes the simple lesson of history as a buttress to skepticism. What is that lesson? That truths held sacred in one period have been emphatically or casually abandoned in another; that human views have changed beyond recognition over the years, let alone millennia. Contemporary fundamentalists often argue that it is only in very modern times that basic truths have been contested, or ideas changed, doctrines overhauled, or habits altered. This is untrue. Although we may be living in a period of accelerated change, these periods of dislocation and abrupt shifts in mores have occurred before; and they will almost certainly happen again. All of human history is an account of vastly different moral standards and practices, changing over time and in space. And the reason for this, Montaigne intuits, is the fallibility of human understanding itself:

> Since a wise man can be mistaken, and a hundred men, and many nations, yes, and human nature is mistaken for many centuries about this or that, what assurance have we that sometimes it stops being mistaken, and in this century it is not making a mistake?

The answer is that we have no such assurance. And Montaigne was right about his own time: we do indeed look back at burnings at the stake and wonder how anyone, let alone any

Christian, could have justified it. But justify it they did. In my own short lifetime, it has been regarded as a truism in America that interracial marriage is an abomination. Now it is widely regarded as completely unobjectionable. The role of women has similarly been transformed in the West, while it has gone backward in many parts of the Muslim world. Montaigne, mind you, isn't asking us to disparage the past from the comfortable certainty of the present. He is asking us to question all of it, to ask each question anew, to take nothing for granted, to be careful of generalizing, and suspicious of dogma.

It is all very well, he argues, determining that a society should reflect one view of the human good or virtue, but how do we determine what that virtue is? "There is no combat so violent among the philosophers, and so bitter, as that which arises over the question of the sovereign good of man." Here's Montaigne's measured response to the advocates of "natural law":

> They are funny when, to give some certainty to the laws, they say that there are some which are firm, perpetual, and immutable, which they call natural, which are imprinted on the human race by the condition of their being. . . . [T]he only likely sign by which they can argue certain laws to be natural is universality of approval. For what nature had truly ordered for us we would no doubt follow by common consent. And not only every nation, but every individual, would resent the force and violence used on him by anyone who tried to impel him to oppose that law. Let them show me just one law of that sort—I'd like to see it.

So would I. Montaigne sees human disagreement about even basic morality not as some function of multiculturalism or modernity or postmodernity, but as the undeniable fact of the human condition. It has always been thus: "The murder of infants, the murder of fathers, sharing of wives, traffic in robberies, license for all sorts of sensual pleasures, nothing in short is so extreme that it is not accepted by the usage of some nation." In one essay, Montaigne even defends cannibalism, when compared to the religious torture practiced by the fundamentalists of his own day.

Then there is simple emotion. In most people, it colors reason. We get into an argument and immediately our pride and self-esteem are at stake. It is hard to concede error; harder still to acknowledge self-deception, or to take the imaginative leap of seeing things through another's eyes. So we stick to our rhetorical guns; and sometimes forget that they are merely rhetorical. We can talk ourselves into things. We can sincerely make arguments that rest on a premise we haven't yet considered fully. We can get corralled by group-think or peer pressure. Or we can be momentarily dazzled by a politician's speech or a filmmaker's skill and fail to see the logical fallacies they're purveying. Or we can be traumatized by an event and see the world through the prism of fear rather than sober reality. All these things are possible because we are human. And at any point in time, we never know for sure who is right.

I think of my own analytical errors in the past few years. Looking back, I can see that my outrage at the atrocity of September 11, however merited, may well have blinded me to the intricacies and dangers of a subsequent war in Iraq. I can see the comedy and tragedy of an entire debate almost all of which was premised on what turned out to be a falsehood: that Saddam Hussein had weapons of mass destruction. This falsehood was taken as fact by every major intelligence agency and by both supporters and opponents of a war to depose Saddam. We were all wrong. But, at the time, we were all passionately convinced of our rightness. What we lacked was the humility Montaigne urges upon us, the detachment he managed to achieve, the self-awareness that makes prudence and skepticism our constant companion, rather than an occasional, retroactive acquaintance.

Does this amount to a complete relativism and nihilism? Montaigne's own work suggests the answer. Every essay is a conversation with himself; every paragraph an observation or an argument. These are not the works of a man who has given up on reason; they are simply the works of a man who knows its limits, and explores them relentlessly. Is Montaigne's skepticism self-refuting? Shouldn't skepticism be skeptical about skepticism itself? In which case, aren't we back where we started? The response is embedded in the motto Montaigne inscribed on his coat of arms: "Que Sais-Je?" What do I know? The statement is itself a question. It hangs provisionally in front of us, like our lives. A circle is not self-refuting; it is self-perpetuating. And beware the man who discovers a straight line.

Far from abandoning reason, skepticism is besotted with it. Like Plato's dialogues, Montaigne's essays are themselves an internal dialogue. Neither Socrates nor Montaigne was a nihilist or relativist. "A spirited mind never stops within itself," Montaigne writes. "It is always aspiring and going beyond its strength; it has impulses beyond its powers of achievement, if it does not advance and press forward and stand at bay and clash, it is only half-alive. Its pursuits are boundless and without form; its food is wonder, the chase, ambiguity." Montaigne is curious about everything. He is insatiable in absorbing the latest gossip and the oldest texts, stories from travelers and his own daily observations. The world is a miraculous mystery to him, constantly inviting further inquiry. "Even now truth finds it necessary to stifle her yawns when she is expected to give answers," Nietzsche writes. "In the end, she is a woman; she should not be violated." She should, rather, be courted. Endlessly.

What matters here is not the establishment of permanent, eternal truths, but the maintenance of important distinctions, the marking of nuance, the weighing of things from different perspectives, the desire to understand something as it is, and not as we would like it to be. If the nihilist says that nothing is true, the skeptic proffers something a little different. Everything is true, he seems to say, as long as it is never taken to be anything more than it is.

But what is it? What does it mean? Was it equally true yesterday? Does another see it differently? The nihilist and relativist have no need for these questions. The skeptic lives and breathes them. The point is to get things right, to relate our own thoughts and therefore words to the reality we are attempting fallibly to describe. This is an arduous and never-ending process—but it is the essence of a fully lived human life. It requires care and precision and alertness to detail, as well as a great deal of nerve. "Truth," Hobbes wrote, "consisteth in the right ordering of names in our affirmations." And reason is integral to making that ordering right.

This is not the arrogant Reason of the Enlightenment, hailed as the light that will eventually banish all darkness born of prejudice, superstition, or power. We have seen the consequences of such folly. Some of the great evils of the past century were purported to have been based on Reason. Marx wrote with the conviction of someone who had proven something empirically true about history; his predictions of a future communist revolution were, in his eyes, scientifically accurate and the summation of a lifetime of factual, rational research. The "Truth" he claimed to have discovered was, in fact, a siren beckoning millions toward hell. The same can be said for the grand new Truths proclaimed by the French revolutionaries. The moment you are told that an infallible Truth is about to set mankind free, it is only a matter of time before the guillotines and gulags are constructed. Only such Truth can justify such acts of barbarism and cruelty to enforce it. And those who question it are not conversationalists, welcome to the postprandial table. They are heretics, herded to the flames.

The lowercase "reason" that Montaigne embraced and that Plato exemplified is a much more modest form. It is the reason we employ in everyday speech, the reason that allows us to have meaningful arguments, the reason that begins with regular conversation and socratically leads to greater and greater personal enlightenment.

The skeptic is also profoundly aware that there are different discourses within this kind of reason and that human civilization has developed sophisticated modes of understanding different things in different ways. An oil painting, for example, may be viewed scientifically as a conglomeration of chemicals and substances; it can be understood historically as the product of a particular time and place and artist; it can be viewed practically as something that we can buy now in order to sell later at a profit; it can be judged aesthetically, depending on its visual power and artistry; if it is designed to promote an argument, to shock or provoke, it can even be viewed politically; if it is an image of the Prophet Muhammad, it could be seen as idolatry. Which one of these viewpoints is the true one? Are they not all true, from the perspective of each discourse?

There is, in other words, no single linear truth about anything, merely perspectives whose particular truth can be judged or measured. It may be true that the painting is in oils; we may be able scientifically to isolate its precise chemical formulation. But that will not tell us the whole truth. Ultimately we may decide that its core truth is an aesthetic one: that a painting should be judged primarily as a work of art—to contemplate, marvel at, be entranced by. But this primary truth is always open to further questioning. And the scientist and the historian are not wrong as such. They are right as far as they go.

For modern human beings, moreover, these distinctions are readily acknowledged. Very few of us want to have an appendix removed by someone versed in art history, rather than in medical science. We do not regard the truth of a painting's price value as somehow an assault on its aesthetic worth. We accept that there are different spheres of expertise, and that one does not necessarily contradict the other. But where we do get into conflict is in the intersection of philosophical and religious truths with practical life. That, after all, is precisely the nub of the argument between the conservative and the fundamentalist. The conservative insists on a clear distinction between what is true from the viewpoint of eternity and what is true from the viewpoint of acting and choosing in the here and now. The fundamentalist insists that such a gap be closed, that the distance between heaven and earth must be erased, if not now, then after death, and if not after death, then at the end of time. The conservative views this as a hopeless and self-defeating illusion, a misunderstanding, and a dangerous invitation to extremism and zeal. He may be depressed by this imperfect reality, or he may be liberated by it. But he is clear that it exists.

III

MONTAIGNE INSISTED ON THIS FROM THE PERSPECTIVE OF radical skepticism about everything. It was his intellectual and existential humility that prevented him from proclaiming that something true forever must dictate every action on earth. But the conservative, while being informed by this radical skepticism, need not go as far as Montaigne. He can simply argue that the truth that applies to eternity is simply a greater and different truth than that which applies to acting in the here and now. He can content himself by observing that the very kind of reason necessary for practical life is simply different in kind than that invoked by philosophy or revelation. The difference between the eternal truth and the truth that governs our everyday lives is as stark as the difference between aesthetic truth and scientific truth. Confusing them muddies the purity of philosophy and theology with the necessary compromises of practical life; and it has come to undermine and even destroy our capacity to live fully and happily as humans.

I know of no writer who has explored this predicament as thoroughly or as insightfully as the twentieth-century British political philosopher Michael Oakeshott. He is, in my judgment, the most important philosopher of conservative thought since Edmund Burke, and he's only just beginning to be given the attention his work so richly deserves. Reading him altered my own life, and helped explicate its conflicts and contradictions. This book would not exist without his inspiration.

Oakeshott wrote only two major books in his life. The first, published in 1933, was an ambitious, intense attempt to describe all of experience—no less—in the philosophical idiom of Hegel and F. H. Bradley, an English Idealist. He called it *Expe-*

rience and Its Modes, finding in experience different "modes" of
seeing the world: science, history, practical life, and, in a later,
marvelously composed afterthought, poetry. All of experience,
he argued, could be found somewhere in these worlds, if you
looked hard and carefully enough. His second book, *On Human
Conduct,* published in 1975, was an account of one mode: practical life, the world in which most of us spend most of our days.
Between these two tomes, Oakeshott taught—he ran the politics department at the London School of Economics for many
years—and wrote dozens of essays and reviews. His only other
full work published in his lifetime, *A Guide to the Classics,* was
not a disquisition on Romans and Greeks, but a guide to horse
racing. Of his intellectual achievements, the most unchallenged
is his theory of history, which ranks perhaps with Giambattista
Vico's in its originality and scope.

Sounds dry enough, and some of it, frankly, is dry. To me,
however, Oakeshott meant something a little different. I first
came across him in an essay I found while doing some reading
on Hobbes. It's not one of his most accessible, and, to tell the
truth, I didn't understand a good deal of it. But there was something in the assurance of its prose, something about his style,
that stuck somewhere in my mind. The essay was in a collection
of his, *Rationalism in Politics.* A couple of months later I came
across it again in a secondhand bookstore and took it home. I
started reading. Within four years of that discovery, I'd tracked
down and read everything Oakeshott had published, and spent
a year doing little else but trying to puzzle him out. It was about
as painless a way of writing a doctoral dissertation as I could
have imagined.

Toward the end of my labors, I got to meet the man himself. I wrote to ask him whether he would be willing to discuss

his work (mine was only the third dissertation ever written on him), and he wrote back, in his elegant calligraphy, that he'd be glad to. I found my way to his village, Langton Matravers, by train from London and arrived in midmorning, in the cloying November mist of southern England. He was waiting at the garden gate, with a mischievous grin on his face, and hustled me inside. The cottage was made entirely of slate and had no central heating. I discovered only three rooms: a cramped kitchen to one side, a poky guest room, and a living room, turned into a makeshift duplex, with a ladder leading to the second story, which this eighty-nine-year-old clambered up each day to get to bed. The walls were lined with books, in various languages, from Victorian potboiler novels to Hegel's *Phenomenology*. He made me coffee, and a meticulous four-course lunch, and, in front of a coal fire, we talked.

My trouble with Oakeshott had always been the contrast between the grandeur of his early ambition and the lyrical detachment of his later achievement. He started his intellectual life with a grand attempt to perceive the truth about the world, and over the decades had slowly shifted into an elegant digression about the world's texture. His last essay was actually a work of allegorical fiction, as if the struggle to perceive the truth had proved too demanding to continue, even pointless. I raised the issue: Did he share the philosophical lassitude of so many of his contemporaries? He replied by saying that giving up the search for "the truth" was not equivalent to believing that it did not exist. His restraint came from modesty, not nihilism. The nuances were everything, perhaps more thoroughly within our reach, perhaps the only way ultimately to understand the whole.

This radical acceptance of what we cannot know for sure is what he put at the heart of his idea of the conservative temperament, and it is why many modern, and especially American, conservatives find him so difficult a figure. This disposition is alien to them: it is fickle, aloof, humane, where they are consistent, engaged, and rationalist. Oakeshott couldn't care less about politics as such, who wins and loses, what is now vulgarly called the "battle of ideas." He cared about understanding the relations between human beings, and he saw the vagaries of people as occasions for celebration, rather than correction. His paradigm was dramatic, not programmatic. His life was poetic, not prosaic. His conservative politics were not a means to repress or regulate man's often sinful exuberance, but a way to allow it to flourish when politics ends.

He saw the way human beings actually live their lives, and marveled at their capacity to navigate the daily grind through the resource one might call practical wisdom—prudence, judgment. He argued passionately that no theory could ever explain this, no book ever fully describe it, no abstraction ever capture it. And like many others, he expressed this truth in metaphors and stories as well as arguments.

Here's an example he often deployed. A great chef can write a cookbook, he would argue, and in that cookbook, she can provide every ingredient, every technique, every precise measurement and time necessary to create a great dish. But the dish that you or I will make from her instructions will never and can never approximate the dish as she could make it herself. Her skill has been acquired through doing; and there is no shortcut to her skill, her accumulated wisdom, her feel for the right ingredient at the right time. The greatest chefs may even reach a point where they have no need for a cookbook or even measurements. They know their way by habit and training and experience and judgment. Their taste has been honed through time;

their ability to cook eludes any attempt to abstract from it. The very quality that no words can capture and no theory explain is the missing ingredient in our modern understanding of human life.

In a footnote, Oakeshott tells a story to make his point. With his customary eclecticism, Oakeshott turns to a tale from Taoist literature to illustrate it. He recites the story of Duke Huan of Ch'i who is reading a book of philosophy at one end of a great hall. At the other end is a mere wheelwright, manually carving perfectly round, wooden wheels. The wheelwright asks if the great thinkers who wrote the book are still alive. "Oh, no," said the Duke, "they are dead." "In that case," said the wheelwright, "what you are reading can be nothing but the lees and scum of bygone men." The Duke, enraged by the man's impertinence, threatens to have him executed if he cannot come up with a good reason for saying what he just said. And this is the wheelwright's reply:

> Speaking as a wheelwright, I look at the matter this way; when I am making a wheel, if my stroke is too slow, then it bites deep but is not steady; if my stroke is too fast, then it is steady but it does not go deep. The right pace, neither slow nor fast, cannot get into the hand unless it comes from the heart. It is a thing that cannot be put into words; there is an art in it I cannot explain to my son. That is why it is impossible for me to let him take over my work, and here I am at the age of seventy still making wheels. In my opinion it must have been the same with the men of old. All that was worth handing on, died with them; the rest they put in their books.

The skill the wheelwright developed he could only acquire in the doing. This skill is what fundamentalism of all sorts, secular and religious, fails to understand. The wheelwright doesn't learn what to do from a doctrine; he learns only by doing it. And the breakthroughs and changes in his practice come from these endlessly accumulating acts of doing, acts that create skill or judgment that in many ways are as ineffable to our abstract rational minds as God is. All practical activity partakes of this; that's why experience counts. And if all practical activity is like this, it follows that the same is true of moral life, because moral life is about the way we act, not the way we think.

There is no way, Oakeshott argues, to generate a personal moral life from a book, a text, a theory. We live the way we have grown accustomed to live. Our morality is like a language we have learned and deploy in every new instant. Indeed, language is a metaphor that Oakeshott returns to again and again. Do we learn language from a book? Of course we don't: "There is no point in a child's life at which he can be said to begin to learn the language which is habitually spoken in his hearing; and there is no point in his life at which he can be said to begin to learn habits of behaviour from the people constantly about him."

It is, of course, possible to infer a dictionary and a grammar from the way people speak at any given time. And the fundamentalist linguist will try to enforce the rules and words in the book on the people who speak the language. He will correct grammar, order sentences, prohibit words, instruct on usage. In France they even have official bodies that rule on whether something is indeed French or not. And in some sense, all of the English language can be contained in a dictionary.

But the dictionary ultimately misses the point. It is merely a snapshot of a moving image. The moving image is language as it is spoken. And there are some among us who have a gift for speaking and writing, whose fluency and imagination and inventiveness remakes the language that we all speak, in ways that no dictionary will ever or could ever capture. To hear the recording of a speech by Churchill or a brilliant peroration in a Baptist church, or the argot of a street corner is to witness something else, more elusive and more beautiful: a human moment when all the rules are temporarily suspended and new rules created merely by the doing, the speaking. There are rules for basketball, but none of them will ever approximate the ineffable movement of a Michael Jordan in his prime. There are even rough and ready rules for jazz, the most purely conversational and therefore conservative of musical forms. But nothing can ever capture hearing a skilled jazz musician improvise, create new horizons and new possibilities in front of us, as we listen. That's when something magical happens.

Oakeshott argues that this is as true of our moral lives as well. The fundamentalist refers to a book or a doctrine and then attempts to apply that ruthlessly to his entire life. He fails, of course. And these moments of failure, of sin, lead to renewed, ever more aggressive efforts to conform his actions to the dictates of the book or the doctrine. The conservative's moral life, in contrast, is far less self-conscious and spasmodic. It is acquired from those he grew up with and subsequently from those he gathers around him. It develops over time with its own rhythm and cadence, it is informed by moral education and primarily understood by observing the examples of others. Sure, we read the texts, we listen to the sermons, we try to apply them to our lives. And none of this is necessarily to be rejected as part of moral activity. But our real morality comes when we have put these guides behind, and have developed a way of living that simply integrates these lessons into an unself-conscious whole.

We do not and cannot acquire our moral lives by books or words alone. We come across people whose lives we find admirable, and we watch them. We deal with them; we are loved by them; and, over time, if their example continues to impress and inspire, we cannot help ourselves in quietly adjusting our own lives to match aspects of theirs. This is how we grow and learn. These habits of moral life are immune to definition or description; they can merely be observed, the way one might look over the shoulder as a painter wields a brush.

In this sense, our religion, our moral life, is simply what we do. It is how we are. It cannot be reduced to a doctrine or a book, although it may find intermittent or even continual inspiration from both. A Christian, in other words, is not a Christian simply because he agrees to conform his life to some set of external principles or dogmas; or because at one particular moment in his life, he experienced a rupture and changed himself entirely. He is a Christian primarily because he acts like one. He loves and forgives; he listens and prays; he contemplates and befriends; his faith and his life fuse into an unself-conscious unity that both affirms a tradition of moral life and yet also makes it his own.

The fundamentalist is always positing an external moral ideal that he must necessarily fail to attain—because he is human. It is outside himself and his job is to internalize it. Fundamentalism as a way of life is therefore a series of ruptures and reforms. It is a cycle of attempts to conform to an external, eternal ideal, and to repeat the process of sin, redemption and sin again indefinitely. In Oakeshott's words, "it has a great capacity to resist change, but when that resistance is broken down, what takes place is not change but revolution—rejection and replacement."

For the fundamentalist, a human being's internal compass, what he has absorbed simply by being who he is, is always suspect—because the self is sinful and must always be subject to correction from the outside. And so the fundamentalist learns to distrust himself, to wrest himself from certain habits, to conform what might have been his personality into a persona that is a vessel for something far greater than himself. Rather than learning from others, he may regard others as moral dangers, who themselves need constant monitoring and correction. And so he does what he can to lose himself in God, or rather in an idea of God, buried in an ancient text, or upheld by an infallible pope, imposed by other humans, and monitored by them.

This kind of moral life, as we have seen, is often marked by the need to deny when one fails to conform to the ideal, or to punish oneself aggressively for waywardness. It zigs and zags from purity to sin and back again. Its motor is guilt, its achievement is often self-loathing, mitigated only by the faint hope of divine forgiveness. Every now and again, one particular doctrine may stand out as vital, and become an obsession—and this seems to be particularly true of the sins of sexuality, that Dionysian force of nature that seems constantly to push human beings into places they are taught to abhor but internally cannot resist.

Sex is so often the object of obsessive compulsion and repulsion for the fundamentalist, perhaps because it is one area of human life that is very difficult to control and often takes place in private or the shadows, away from public scrutiny. And so the fundamentalist has to make an extra effort to seek it out and punish it; he will be particularly concerned to restrain it by law or custom or social pressure; or he will insist that it be reduced to a single purpose—procreation—rather than accepting it as just one aspect of being human that can be expressed in a variety of moral and immoral ways. And he will be very leery of believing that the intense and sublime pleasure and play that lie in sexuality can actually be a way to intuit the divine, rather than an invitation to surrender to Satan.

In the most extreme cases, certain forms of sexual repression can become a new form of god in themselves, a pivotal criterion by which to judge the entire moral core of a person, a shortcut to assessing virtue or vice as a whole. And so we lose perspective. And the chaste Christian discovers one day that his obsession with sexual purity has also led him to be callous to his family or mean to his colleagues or self-important among other Christians. Or the natural lawyer, determined that his truth be realized for all mankind, finds himself supporting laws that would send policemen into bedrooms, and doctors into jail. He never meant to be cruel, but his faith demands it. Salvation requires it. Oakeshott again:

> Too often the excessive pursuit of one ideal leads to the exclusion of others, perhaps all others; in our eagerness to realise justice we come to forget charity, and a passion for righteousness has made many a man hard and merciless. There is, indeed, no ideal the pursuit of which will not lead to disillusion; chagrin waits at the end for all who take this path.

When individuals try to govern themselves this way, they can experience great bliss at times, but crushing disappointment at others. The fundamentalist, in seeking perfection, may find himself farther and farther from the God he seeks and the virtue he craves. Why? Because we cannot *think* ourselves into godliness. And those who construct rules and regulations for every minutia of human activity, who have an answer to every moral conundrum, an authoritative solution to every dilemma, cannot solve this recurring predicament. Relating what is true forever to what we do here and now is always a crooked line, like a fishing line into a lake. Every attempt to insist on its straightness will founder, in the practical world, on the different elements of air and water.

A friend of mine told me in an e-mail how she tried to make the crooked line straight, and eventually realized her mistake: "Coming out of fundamentalism myself, I can especially relate to the endless cycle of guilt, repentance and moral striving that goes nowhere. Only when I came to the end of myself did I finally allow the life of Christ residing within me to do its transforming work. One realization I came to was, instead of viewing sin and failure as something to be overcome at all costs, I now view it as the very means by which I can draw nearer to God and to other people. It keeps me humble. Maybe that's the path Jesus intended us to take when he taught about love. I don't think some fundamentalists understand the paradox of the Christian life, how Christ's incarnation, death, and resurrection have transformed the way we relate to everyone and everything. Many fundamentalists view these things as outsiders, but they do not comprehend or participate in its mystery. I think that's why they live by law. They can only go through a series of empty motions that are a poor imitation of living by faith."

The message of the Gospels seems to me to be constantly returning to this theme: those who set themselves up as arbiters of moral correctness, the men of the book, the Pharisees, are often the farthest from God. Rules can only go so far; love does the rest. And the rest is by far the most important part. Jesus of Nazareth constantly tells his fellow human beings to let go of law and let love happen: to let go of the pursuit of certainty, to let go of possessions, to let go of pride, to let go of reputation and ambition, to let go also of obsessing about laws and doctrines. This letting go is what the fundamentalist fears the most. To him, it implies chaos, disorder, anarchy. To Jesus, it is the beginning of wisdom, and the prerequisite of love.

My favorite of all the stories told about Jesus is one of the simplest. At one point in my life, when I was diagnosed with what was then a fatal disease, HIV, when one of my closest friends was suddenly admitted to the hospital with AIDS, and when my mother was also hospitalized with depression, I felt something inside me simply beg for God's help. I wanted to know why all these things had befallen me and those I loved all at once. I wanted an answer. I wanted something to hold on to, something to anchor me, to return to me the spiritual and physical equilibrium I had suddenly lost. I found myself drawn to the Gospels, and all I can say is that the old story I had long loved spoke to me more powerfully than ever.

The story is of Jesus's surprise visit with two friends, Martha and Mary. When Jesus arrives, Martha immediately does what she should: she prepares food. The meal is in the future and her job is to get there. Mary, in contrast, simply "sat down at the Lord's feet and listened to him speaking." Martha gets progressively more irritated with Mary's indolence and finally bursts out: "Lord, do you not care that my sister is leaving me to do the serving all by myself? Please tell her to help me." Jesus answers: "Martha, Martha, you worry and fret about so many things, and yet few are needed, indeed only one. It is Mary who has chosen the better part; it is not to be taken away from her."

An endorsement of idleness? Of irresponsibility? Of selfishness? In a way, Mary is guilty of all these moral failings. By the book, she's wrong. But in that very moment, she is not merely right. She is, in Jesus's formulation, doing the *only* thing that is right. And she is doing nothing. She is merely being with Jesus. She has let go.

And this, it seems to me, is the true mystery of the incarnation, the notion that in Jesus, God became man. I should say that I believe this in the only way I can: that one man represents, for all time, God's decision to truly be with us. The reason I call myself a Christian is not because I manage to subscribe, at any given moment, to all the truths that the hierarchy of my church insists I believe in, let alone because I am a good person or a "good Catholic." I call myself a Christian because I believe that, in a way I cannot fully understand, the force behind everything decided to prove itself benign by becoming us, and being with us. And as soon as people grasped what had happened, what was happening, the world changed forever. The Gospels—all of them, including some that were rejected by the early Church— are mere sketches of a life actually lived, and an experience that can never be reduced to words or texts or doctrines. And the world as it was—as it still is—was unable to tolerate this immense occasion; and so Jesus was executed and the life more in touch with divinity than any other life was ended abruptly, when it was still achingly young. The existence of such a life was both so wondrous that it changed everything; and also so terrifying it had to be snuffed out.

The point of this incarnation was surely not to construct a litany of offenses by which we are to judge our own lives at any moment, to force us to thrash and writhe in a constant ordeal of self-criticism and guilt. The point was merely to be with us; and by being with us, to show us better how to be human, how better to embrace our lives by accepting the divine around us

THE CONSERVATIVE SOUL · 209

and inside us. By letting go, we become. By giving up, we gain. And we learn how to live—now, which is the only time that matters.

IV

IN THIS NONFUNDAMENTALIST UNDERSTANDING OF FAITH, practice is more important then theory, love more important than law, and mystery is seen as an insight into truth rather than an obstacle. This is the Christianity that the conservative clings to; and it is a form of Christianity the fundamentalist rejects. That is his right and his prerogative. But it is the great lie of our time that all religious faith has to be fundamentalist to be valid. There is another way. For Christians, that other way is about a man, Jesus, whose individuality and humanity cannot be abstracted. And it is about a commemoration of that man, as he asked us to commemorate him—in a meal, a breaking of bread, a Seder-made-new, the mass, as Catholics have come to understand it. This is my faith, if I were forced to describe it. It is not about doctrine as such, although doctrines may indeed be inferred from it. It is about a man who was God and lived among us, and died for us, and gave us his own body and blood to eat and drink, and who commanded us to ignore everything except the only thing necessary: the command to love one another, as a sign and reflection of God's love for us. I am by no means perfect or even good. But this faith has never left me; and I refuse to acquiesce to the idea that it isn't as real as any fundamentalist's.

This is a faith that rejects the eschatology of fundamentalism, that sees the present not as a burden before a future salvation, but the only place our souls ever truly live. The fundamentalist, as we have seen, finds time a difficult but superable concept. We may live imperfectly now, but we will one day live differently, he asserts. Life on earth is inevitably inferior and prior to life in heaven. Salvation beckons with urgency. And so, in many ways, the fundamentalist, like Martha, lives constantly in the future—of his own life after death, and of the world's ultimate ending.

A core feature of fundamentalism is the notion that the End-Time will come, and the awful, unbearable disjunction between earth and heaven will some day be erased forever. And a core tenet of the fundamentalist way of life is suppressing our human failing now in order to receive our reward of eternity later. History, moreover, has a direction. It has a meaning. The chronology is that of the Fall, the redemption won by Christ's atonement for our sins, his resurrection (which is a necessary proof of his divinity), our eventual deaths and salvation, and ultimately, the salvation or damnation of all. The world is, at any moment, at some point in this process. Providence guides it. Christ, in some ways, is a logical necessity for it all to work.

I'm not trying to argue here that this is somehow wrong. How could I? It makes internal sense, it provides a narrative by which we can attach meaning to the universe, it is clearly part of the Gospel narrative, and has been central to the teachings of most Christians for millennia. Chronology seems necessary to make sense of a temporal world; and it intimates something we call progress. But what if time is actually unlike that? What if eternity is unlike that? What if the eschatology created by those early Christians expecting the imminent return of Jesus is as mistaken in theory as we now know it was in practice?

Oakeshott proffers a different notion of what time means and what history is. The most essential quality of human history, he argues, is contingency. By contingency, he means simply that one moment may or may not lead to another. History is not a process, an unfolding of a single, coherent narrative. It is, rather, a series of countless human events; it is a story of decisions made to act, of choices which can go either way; and could have gone either way—or many ways, none of which can be safely predicted in advance. It is a story of human freedom, of choice in the context of ultimate uncertainty. It has no direction.

A Marxist will counter that the real meaning of history is the unfolding of class conflict, of materialist economic forces leading to an inevitable revolution. A Whig—or the twenty-first-century version of a Whig, a neoconservative—will describe it as the gradual and unstoppable advance of human freedom. A Hegelian may argue that it is the result of a philosophical dialectic in human self-understanding. A Christian fundamentalist will explain that it is a series of dispensations leading inexorably to the Final Judgment. A secular liberal will see it as the slow victory of good ideas over bad ones.

The conservative, in contrast, will argue that we cannot actually know if there is some meta-narrative to our human past. We can only know from our own lives, by an awareness of our own paths, and by a careful study of our collective pasts, that history is necessarily a dynamic succession of uniquely and distinctly chosen actions, each of which has incalculable and often unintended consequences. You make one decision and suddenly a whole variety of new choices beckon, many of which have been occasioned by your most recent call. Each moment, each choice, refashions not just what is ahead of us, but also gives fresh perspective to what has happened to us in the past. In T. S. Eliot's words again,

the pattern is new in every moment
And every moment is a new and shocking
Valuation of all we have been.

History is the total sum of all these countless decisions, some of which, obviously, affect more people than others, but all of which change the world in some way or other.

Think of history, in other words, as a gigantic pool game. Each strike, each individual human choice, transforms the table. The only difference is that, in this pool game, there are no pockets for the balls, no ultimate winner. It is a dynamic, dramatic game that continues forever; its only rule is that a ball can be struck at any angle with any number of possible consequences. It is an astonishing interplay of unknowables. It can lead anywhere, and ultimately goes nowhere.

Tradition, on this reading, is simply the pattern that exists at any given moment on the pool table. It is where you start from, it constrains what you can do, it commands attention and respect, and yet there is still enormous potential for change. A skilled player will immediately intuit imaginative ways to reorder the whole table, or to play it safe, or to just move it along. In Oakeshott's words, "A tradition is not something to which we must adhere; it is something which provides the starting point and the initiative for fresh enquiry. It is no use looking to it for finished conclusions, for settled answers to fixed questions because it is not a tradition of conclusions or even of questions, but of enquiry."

This is what time is, he argues, and it is the universe in which practical life has to occur. One thing leads to another, and every moment presents us with choices of how to act and what to do. Yes, there are constraints: the historically contingent pattern you are born into; the genetic lottery; the hazards of physical life. But in the end, practical life does not relent in offering every individual a constant array of choices, trivial and profound, that she has to make. Even not making a decision is a decision.

There is no escaping this freedom, and it can paralyze the truly thoughtful, energize the unscrupulous, and entice the opportunist. The most adept of human beings master their practical environments. They are brilliant at deploying words and acts to achieve what they want. These are the shrewd businessmen or investors, intuiting the next market niche, persuading their fellows to part with their money, getting out of a stock at its peak, and creating wealth that enhances their power, which in turn, if deployed with finesse, can lead to more wealth.

Machiavelli was peerless in explaining how humans can exploit the world, chance, and, above all, each other. Forget God, he seems to say. Just get on with making your way. The most successful Machiavellians, mind you, would never say "Forget God" out loud or in public, because that might impede their aims, disillusion those they need to marshal, or subject them to unnecessary opposition. And some of the most practical of practical men have been religious leaders, and still are. Others, in turn, have even manipulated the contingencies of life and history for what seem to be good and decent ends. These are those we might call statesmen. They are men of the world, but they are not out solely for their own advantage.

The conservative, unlike the fundamentalist or Marxist or any other adherent of a direction for time, simply observes that this is the way the world is. He will confront the fundamentalist with a puzzled look, and ask him how he knows for sure that something beyond contingency and choice is at work in human history, that some other force is directing human action and ends. He will note, as Thomas Cromwell does to Thomas More in Robert Bolt's *A Man for All Seasons,* that along with prayer, there is also something called effort. He will enjoy pointing out the collapse of this great theory of history and that one. And in the meantime, he will simply make the choices he wants to make and live. Laurence Olivier put the conservative temperament in this respect rather well when he said: "I take a simple view of life: keep your eyes open and get on with it." Call it an acceptance of presentness.

The conservative will therefore tend not to be surprised by the way things turn out, because he is quite prepared for any eventuality. Some describe conservatives as pessimists because of this. Conservatives, they say, are rarely disappointed because they rarely hope. But this is simply an extreme version of the conservative outlook. Another kind of conservative is quite prepared to be surprised by an unexpected turn for the good, as much as for the bad. Some conservatives may even be temperamentally cheerful, while also aware that bad things happen and that humans can make terrible decisions about their own lives and those of others.

The point about contingency, after all, is that it is contingent: anything can happen. Conservatives know that people can always surprise you; inevitable catastrophes sometimes don't actually happen; unforeseen disasters can also be right around the corner. The key to being human is knowing that chance and choice determine our lives, and, rather than resisting that, rather than insisting that he can know and attest to a deeper meaning or direction, a conservative will simply accept the limits of his own practical knowledge.

We see the serendipity of practical life in our own lives and in our collective pasts. We remember a moment when we stumbled across the person we end up spending the rest of our lives with. We didn't intend to. We were looking for something else entirely. But there he is. A chance phone call, like thousands of others, turns out to be the one that gives you a new career, or takes you to a new place, that, in turn, opens up a new world of choices. For decades, many Americans expected that they might perish in a sudden nuclear Armageddon in a conflict with the Soviet Union. The Communist system appeared as an invincible and malign behemoth, slumped across half the globe. And then, with a push and a smile, it collapsed—with the help of a former B-movie actor, a British female chemist, and a playwright-priest from Poland. And when war did come to America, it happened literally out of the blue, as a plane pierced a tower and ended an era. This is what our lives are like. This is what life is: unpredictable, dynamic, uncontainable.

The fundamentalist sees in Darwin a threat because he proffers an alternative version of history, natural and human, than is written in their sacred texts; the conservative sees in him an interesting ally. What Darwin discovered was that our very bodies are the accidental products of countless millennia of random selection—yes: random. The shuffling of the genetic pool has no direction in itself. And evolution is, in some ways, a misnomer, because we want to think of that word as somehow implying progress, as having a direction, a meaning in itself. But it doesn't. Species evolve purely in response to changing environments in an effort merely to survive and propagate. What emerges is incidental to this process, and those who are in the midst of it have no inkling of what lies beyond them.

Humans flatter themselves by regarding their species as the most self-conscious and therefore the highest. This is simply another projection. We are not the most successful life-form on earth; we are not the oldest; we are fragile compared to some. Our environment constantly changes, and so do we. From the point of view of mere survival, we are no more successful than the astonishing life-forms that have recently been discovered in the darkest depths of the ocean. In this sense, social Darwinism is the inverse of real Darwinism. There is no purpose here but survival; no moral lesson to be gleaned from it; no intelligent design but that dictated by the impulse to life itself. There is simply constant change, continuous adjustment, and random selection in an eternal present. A fundamentalist is shocked by this; a conservative knew it all along.

The fundamentalist looks at the diversity in humanity and all of nature and sees "objective disorders" all around him; he needs to impose a system on it all, to give it all a purpose and a direction. There is one correct way to be human, he insists. Once we have discovered it, it is our duty to encourage and impose it. Insofar as Darwin undermines the ideology behind this logic, Darwin is dangerous. The conservative, in contrast, is fascinated by the little and big differences; he accepts the diversity as an empirical reality and subsequently attempts to understand it better. It doesn't threaten him because he too is a part of it. When there is so much to marvel at, his existential loneliness yields to wonder. He finds himself saying, as Gerard Manley Hopkins expressed it,

> All things counter, original, spare, strange;
> Whatever is fickle, freckled (who knows how?)
> With swift, slow; sweet, sour; adazzle, dim;
> He fathers-forth whose beauty is past change:
> Praise him.

V

THERE IS NO MISTAKING, HOWEVER, THAT LIVING IN A universe that provides no direction, that offers no ready, externally provided answers to our deepest questions, can be an unnerving place. How do we come to terms with this? How do we live with this? We can reach for theories to give what has no direction a destination; we can revere books that purport to answer the randomness of fate; we can deny reality. But these strategies lead to inevitable failure. And insofar as none of us can live entirely without some recourse to these books and theories, we have to live with that failure. But how do we make peace with it?

This, Oakeshott argues, is where religion, in its most perfect sense, comes in. One of Oakeshott's metaphors for the practical life, for the life we all live, is of water. Water is an element that can both calm and terrify, like religion. In the shared world we live in,

> men sail a boundless and bottomless sea; there is neither harbour for shelter nor floor for anchorage, neither a starting-point nor appointed destination. The enterprise is to keep afloat on an even keel; the sea is both friend and enemy; and the seamanship consists in using the resources of a traditional manner of behavior in order to make a friend of every hostile occasion.

The fundamentalist will insist that the womb is where we start from and heaven is our destination. He will deduce theories of morality from biology and philosophy, from our natural lives and supernatural futures, and proclaim them true and dispositive. He sees a beginning and an end, and insists that there is one right way to get from one place to another. He has a map, as well, whether it be the Book of Revelation or the Communist Manifesto.

The conservative, in contrast, is not so sure. The books are interesting, but real life requires that they be occasionally or often put to one side. More to the point, he sees the horizon constantly disappearing before him. He does not remember his own conception or even birth; his consciousness cannot be easily reduced to a physical formula; he has a mind that is both part of and yet also able to distinguish itself from his body. He becomes who he is and doesn't know entirely what he will be in the future. He will look around him for guidance, and part of that guidance will come from the collective past into which he is born, a past that has often distilled itself into books or institutions or customs to which the conservative will accustom himself. But he will also occasionally challenge these inheritances if circumstances seem to demand it; he will see in them opportunities for new self-expression and new self-discovery. Time, like the sea, is both friend and enemy. And you never know which until the very moment it is upon you.

"[W]ith every thought and action," Oakeshott writes, "a human being lets go a mooring and puts out to sea on a self-chosen but largely unforeseen course. It has no pre-ordained destination: there is no substantive perfect man or human life upon which he may model his conduct. It is a predicament, not a journey."

Is this predicament depressing? Only to those who require certainty to live. But the key to living, Oakeshott suggests, to living happily, is an acceptance of what we temporarily are, even an enjoyment of the lack of finality and the lack of perfection. Faith rests on doubt, not certainty.

Oakeshott's very first essays concerned religious truth, but after the 1930s there had been an utter silence. Absolute, untroubled faith, one could glean from between the lines of his writing, was too acute, even too crushing, an obligation to sustain; or too important, too mysterious, to write about. Oakeshott seemed even to enjoy the ambiguity of half-belief, seeing sin as the occasion of a fascinating conversation with oneself and with God, rather than as an oppressive encumbrance to happiness. "After all," he quipped, half seriously, to me, "who would want to be *saved*?" God might even prefer us as we are: we're more interesting flawed, and, without flaws, no real love is possible anyway, either between us, or between us and God. Love, Oakeshott explained, has its origins in mutual amusement, and ends in "total acceptance" of the other person. Or, as he put it in one of his essays, "Friends are not concerned with what can be made of one another, but only with the enjoyment of one another; and the condition of enjoyment is a ready acceptance of what is and the absence of any desire to change or improve." That, he seemed to say, must be the core of the love Jesus incarnated, the way God loves us entirely, as we are. "It is Mary who has chosen the better part; it is not to be taken away from her."

If the acceptance and love of others as they are is the essence of Christianity, then the acceptance of our loneliness and doubt in a world far beyond our understanding is the core of all non-fundamentalist religion. Some activities seem to speak to this more than others: the activities in life that themselves reflect in miniature what the whole of our existence really is. Fishing, for example. No, I'm not kidding. "Fishing," Oakeshott explains, "is an activity that may be engaged in, not for the profit of the catch, but for its own sake; and the fisherman may return home in the evening not less content for being empty-handed. Where this is so, the activity has become a ritual and a conservative disposition is appropriate."

Ritual is the key word here. Ritual is threatening to those who press living to a point, because ritual has no point but beyond itself. A ritual fully realized is utterly unself-conscious. By doing something again and again, by taking care to observe its forms, by starting on a journey that we know we will only repeat endlessly: this somehow abates our sense of existential lostness. This is why ritual is so intrinsic to religion, because only ritual can approximate the ineffability of the divine, enact its truth while not purporting to explain or capture it. In some ways, this is the conservative account of Christianity. It is first and foremost a single life: of one man, Jesus. It is subsequently both an attempt to distill what it meant to be Jesus—most fatally in the abstract religious genius of Paul—and then a reenactment of his self-giving in a ritual Catholics call the Mass. Finally, it is the absorption of Jesus's way of life into one's own; and the most successful form of this will be one that eventually becomes Jesus-like unself-consciously.

Does Jesus live in the present? In so much as the memory of his life and beauty of his message penetrates us, yes. There are also times when I have clearly sensed the presence of Jesus in my midst and in my soul. I accept that as a mystery I cannot understand. In the same way, I accept the reenactment of his scandalous last supper as a way to incarnate Jesus again—as food. This sacramental version of faith asserts that mystery is with us physically, that God is in everything, and that therefore Jesus is in everything and in everyone. The Mass expresses this—and communion, the literal taking in of Jesus into one's own body, is the fusion of spirit with flesh, a mystery that both ritualizes and expresses the revelation that God chose to be with us.

THE CONSERVATIVE SOUL · 221

It is important that those of us who are Catholics understand the meaning of the Mass. But it is also important that its meaning always escape us. And so there comes a point at which words do not clarify; they obscure. I never knew the Latin Mass and so I can only dream of the ritual beauty the church once had. But even in its sparser, less ornate, and less mysterious form, the Mass is still a physical reenactment of Jesus's self-sacrifice, a meal ritualized and turned into action. Eventually, even the words are acts. I have attended Mass hundreds and hundreds of times in my life, and there are many parts of it that I know, as we say, "by heart." And I have come to realize that those moments of unthinking are not somehow an abandonment of faith, but in some ways its most perfect expression. This, I think, is what Oakeshott was getting at.

Trying to get an abstract meaning of the Mass misses the point. The point is the activity, not the idea. In the sane words of the seventeenth-century divine Jeremy Taylor,

> Is it not enough for me to believe the words of Christ, saying, *This is my body*? And cannot I take it thankfully, and believe it heartily, and confess it joyfully: but I must pry into the secret and examine it by the rules of Aristotle and Porphyry and find out the nature and the undiscernible philosophy of the manner of its change and torment my own brains, and distract my heart, and torment my Bretheren, and lose my charity, and hazard the loss of all the benefits intended to me, by the Holy Body; because I break those few words into more questions than the holy bread is into particles to be eaten?

I remember my grandmother. She was an Irish immigrant to London in the 1930s and worked as a servant and cleaning lady for priests. In her later years, she lived with us and we would sometimes go to Mass together. She was barely literate, the seventh of thirteen children, no intellectual by any means. And she could rattle off the "Hail Mary" with the speed and subtlety of a NASCAR lap. There were times when she embarrassed me—with her broad Irish brogue and reflexive deference to clerical authority. Couldn't she genuflect a little less deeply and pray a little less loudly? And then, as I winced at her Irish volume in my quiet English church, I saw that she was utterly oblivious to those around her. She was someplace else. And there were times when I caught her in the middle of saying the Rosary when she seemed to reach another level altogether—a higher, deeper place than I, with all my education and privilege, had yet reached. Maybe this is partly what Oakeshott had in mind. Sometimes the calmest place is at high speed; sometimes, the securest place is somewhere that, in its actions and rituals, tells us we're home. Not a physical place, but a familiar action.

In religion beauty matters; the aesthetic counts. It counts because the kind of wonder we sometimes feel when looking at a stunning painting or a sublime sunrise is the wonder we should really feel all the time in the presence of God. What religion can be at its most sublime is the fusion of that wonder with practical life. It is the marriage of the poetic and practical modes of experience. This does not require the imposition of fixed rules and doctrines, although they may be helpful guides from time to time. It requires a constant reimagination of the potential of life lived on earth as if it were heaven. It requires letting go of our desire not to let go. Jesus saw it in children. One of his most radical teachings was the notion that only if we become like children will we enter the kingdom of God.

Children love rituals, and their games are full of them. Perhaps because they are not yet fully formed, every new moment matters more. We older types have sometimes become inured to the wonder and mystery of everything. One day among thousands means less than one among hundreds. "Everybody's young days are a dream," Oakeshott remarks, "a delightful insanity, a miraculous confusion of poetry and practical activity in which nothing has a fixed shape and nothing has a fixed price." He loved that; and he loved children. He paid them the highest compliment in his world of calling them "great conversationists." He went even further in elucidating their practical language:

> The young speak it differently from the old, not merely because (as Aristotle remarks) they are less competent linguists, but because what he has to say who "drinks the valiant air of dawn," who is overwhelmed by the limitless invitations of a human existence, and to whom "the long littleness of life" is yet undreamed, calls for the lyric mode. Their ingenuousness is not wisdom, but in their shallow utterance a moral language may acquire a refreshing translucence.

That was what I heard in my grandmother as well: a childlike utterance that reached spiritual transcendence. Listen to the rigid and certain fundamentalist. How childlike is he? How alive is he? For Oakeshott, the religious life "is synonymous with life itself at its fullest, there can be no revival of religion which is not a revival of a more daring and sensitive way of living." Daring and sensitive: an unusual combination, but one that to me reflects the kind of living that Jesus exemplified.

There is something mystical about this kind of religion. It doesn't try to secure itself in a single text. It doesn't obey a single authority. It doesn't go back in time to discover the perfect form or the infallible truth. It partakes of all these things in part.

How could it not? How can a Christian exist without the Gospels? How can a Christian today believe without the church's centuries-long care in protecting an inheritance? How can a Catholic simply ignore the statements of those who have authority and leadership in the institution that baptized and educated him? He can do none of these things, and wouldn't want to. But he will subject all of them to scrutiny and will not stop at any of these points. Such a faith incorporates these things but aims to live them, to translate them into life, and to experience God in the living here and now.

The precious gift of religious life lies in part, in Oakeshott's words, "in the poetic quality, humble or magnificent, of the images, the rites, the observances, and the offerings (the wisp of wheat on the wayside calvary) in which it recalls to us that 'eternity is in love with the productions of time' and invites us to live 'so far as is possible' as an immortal." I'm reminded again of T. S. Eliot's assertion that "to apprehend / The point of intersection of the timeless / With time, is an occupation for the saint." These moments may come upon us when we least expect them. We may see flashes of eternity in the simple grin of a child in a game of hide and seek, in the approach of the tide on an autumn afternoon, in the eyes of a lover in sex, or in a grandmother's ritual—but we know them when we see them. The key is to be open to them, because they happen all the time, all around us. But we are too "busy" to notice.

The opposite of this kind of faith is fundamentalism: the constant recourse to abstraction and authority or text. And its rival in secular terms might be thought of as worldliness, the refusal to see mystery in life, the refusal to see something beyond death, or beyond mere matter or social relations. This can be expressed in a variety of ways: the construction of a life around the approval of others; the pursuit of applause; or of money and fame; the maintenance of power, with no sense of its fleetingness; materialism of all kinds; the fetishization of acquisition, of physical things; the pursuit of worldly power in the full belief that it is all there is.

I remember reading and enjoying Machiavelli, the most hilarious and insightful of all those writers defending worldliness as an end in itself. It is indeed possible to master the world of men and women by deploying force and fraud. You can win and keep winning, if you are shrewd enough. But what peace do you really achieve when it is all over? What composure have you really mastered as you fall asleep at night? The doctrine that sees nothing but this world, that lives now but never sees now as an insight into forever, is a dismal doctrine, and an isolating one. Yes, you may one day, after your death, or even in your lifetime, be celebrated for some sort of achievement. But that is not everything, and it may indeed be a delusion.

Oakeshott wrote about this in an unpublished notebook, retrieved by one of his most insightful students, Paul Franco:

"Achievement" is the "diabolical" element in human life; and the symbol of our vulgarization of human life is our near exclusive concern with achievement. Not scientific thinking, but the "gifts of science"; the motor car, the telephone, radar, getting to the moon, antibiotics, penicillin, telstar, the bomb. Whereas the only human value lies in the adventure and excitement of discovery. Not standing at the top of Everest, but getting there. Not the "conquests" but the battles; not the "victory" but the "play." It is our non-recognition of this, or our rejection of it, which makes our civilization a non-religious civilization. At least, non-Christian: Christianity is the religion of "non-achievement."

If you compare that vision with the comforting bromides of contemporary fundamentalism, the "prosperity Gospel," the mega-churches that look and feel like shopping malls and football stadiums, the obsessive recounting of myriad different texts and passages of scripture as if they themselves were somehow protection against the unappeasable loneliness of being, then you see how far so much of contemporary fundamentalist Christianity has strayed from the religion of "non-achievement." The use of religion for social status is, in other words, a fraud. And the use of religion as a mechanism for social order, as a regulatory scheme to keep human beings in line, or as a unifying principle to herd people to the ballot box, is a profound blasphemy.

A vitally lived Christianity is different. While it absorbs all that has gone before it, it is still alive to reinvention and rebirth. Its essence is not something abstract or in the distant past. To say that it has remained unchanged for two millennia is to say that it no longer exists. "The presentation of [Christianity] must in this respect change to changing needs," Oakeshott argues, "becoming, not for the first time, a religion in which no servile archaeology inhibits vitality or chills imagination. On some of the current theories of the identity of Christianity such a change would imply a destruction of our religion, but I have tried to show that it is these theories which are fallacious and not our religion that is moribund."

This is faith freed from fundamentalism. There are religious options other than fundamentalism—options that reach back to the spirit of Jesus and imbue our lives with a deeper, more vital sense of love and immortality, options that are not afraid of living more fully and deeply in the present, options that seek to explore rather than judge, to wonder rather than to explain, to let go rather than to hold on. This spirit is not absent among many fundamentalists' lives or indeed among "traditional" Catholics. It is too irrepressible for that. But it is impeded by the vines of dogma and judgment that have come to strangle the faith so many still seek.

We live alone and now; and yet we live in a universe without bounds or end. Religion is the form of practical life that best addresses that fundamental human predicament. It exists, in Oakeshott's words, "to satisfy no craving for knowledge apart from the knowledge which comes with the mere strength and courage to take life as it is and 'turn its necessities to glorious gain.'" In his final years, Oakeshott returned more and more to religion, and it is what we discussed before the fire on that dank English November day. As he approached death, it mattered more and more to him. He wanted to write another essay on Christianity, a faith held by Augustine and Montaigne—"the two most remarkable men who have ever lived." He wanted to evoke an understanding of salvation, as he put it in a letter to a friend and scholar, "that has nothing whatever to do with the future."

That essay remained unwritten. But Oakeshott had done something just as valuable. By recasting conservatism as a philosophy that rested on skepticism, not dogma, by rescuing practical life from the aridity of abstraction and ideology, by elucidating an idea of history that defied direction, and by sketching an approach to faith that left it up to each of us to live it as best we can in the present, he gave us an intimation of something else. He helped us understand that the core, animating principle of human life as we now live it is freedom. And securing that freedom is the necessary task of politics.

CHAPTER 6

A Politics of Freedom

"In this present crisis, government is not the solution to our problem; government is the problem. From time to time we've been tempted to believe that society has become too complex to be managed by self-rule, that government by an elite group is superior to government for, by, and of the people. Well, if no one among us is capable of governing himself, then who among us has the capacity to govern someone else?"

—PRESIDENT RONALD REAGAN, *Inaugural Address, 1981*

I

CAN A VIABLE POLITICS BE BASED ON DOUBT? IN TODAY'S fundamentalist world, where religious leaders vie with charismatic conviction-politicians to tell us how to live our lives, it would seem unlikely. There is no more fatal charge for a politician than that of a "flip-flopper." A public figure who changes his mind once is suspect enough; one who constantly adjusts and questions and doubts might as well forget about a successful career. He'd be sliced and diced in a million attack ads before he even got to his first primary.

All of this, however, is an indication of the sickness in our current politics, not its health. It is also a sign of how eclipsed genuine conservatism now is. For conservatism's deepest roots lie in doubt; without doubt, conservatism would have little to offer the modern world. If it were just another ideology, another system of thought vying for public attention and support, it should simply join the queue. There is never any end to those claiming to have discovered the infallible truth, the permanent solution to the human predicament, and a fail-safe way of organizing society so as to perfectly or more accurately reflect this truth. You want a politics that will end all existential alienation? Become a communist. You want a politics that will redistribute wealth and promise social justice and inclusion? Become a socialist. You want a politics that rests its defense of inalienable human rights on a God-given liberty? Become a liberal. You want a politics that affirms divine truth in its governance of human affairs? Visit Iran.

The radical alternative to all these options is conservatism. As a politics, its essence is an acceptance of the unknowability of ultimate truth, an acknowledgment of the distinction between what is true forever and what is true for the here and now, and an embrace of the discrepancy between theoretical and practical knowledge. It is an anti-ideology, a nonprogram, a way of looking at the world whose most perfect expression might be called inactivism.

None of this means, however, that it does not have an argument, or that its argument doesn't lead to a distinctive approach to the necessity of politics. I say necessity because it is simply a fact of human existence that we live together, as humans among humans. There is, *pace* Margaret Thatcher, something called society. The question is simply how we organize or choose not to organize such a society, and what principles lie behind that answer. In the next few pages, I hope to argue for one interpretation—among a variety of possible ones—of what a genuinely conservative politics could look like. I hope to offer a version of what a human being who shares the conservative outlook I described in the last chapter would call politics. And why we need it now as much as we ever have.

The defining characteristic of a conservative, as I defined him in the last chapter, is someone who knows what he doesn't know. And in the context of other human beings, the first obvious deduction from this is that what you don't know can harm you. This was the simple but inspired insight of Thomas Hobbes, arguably the most original political thinker in the history of political philosophy. Hobbes's radical skepticism expressed itself in a metaphor, what he called "the state of nature" of human beings. In a world of doubt, where human beings are physical and fragile entities, you cannot know what the person next to you might do next. You might assume he is a rational, peaceful person. But he may not be. And your trust may be misplaced. In a split second, you could be injured, or stolen from, or even murdered. The starting point of a politics is therefore the establishment of a political order that ensures your physical security as much as humanly possible.

Hobbes argued that there was only one way to do this and that was for one entity to have a monopoly of physical force. That's what most people understand by a government. In most societies there are laws of some sort, rudimentary or elaborate, governing our interaction with one another. But who is to guarantee that these laws are actually enforced? Someone somewhere has to wield the threat of overwhelming violence if necessary—to punish the lawbreakers and keep the peace. And if we are not to degenerate into chaos or civil war, within any given geographical location, that force must be unitary. Most of the time, we are utterly unaware that the threat of massive violence by a unitary sovereign entity underpins everything we do in a free society. But that's a sign of a successful polity. It is one where we have become so used to security that we have forgotten its origins.

We do not need to look very far for the alternative; and recent history has shown us what it is. The difference between Iraq under Saddam Hussein and Iraq after he had been toppled is the difference between a state where massive violence is held by a single actor—the brutal dictator—and one where there is no unitary monopoly of force at all. That force after the toppling of Saddam was supposed to be the American military. But it was not sufficiently numerous to provide order or stability and so Hobbes's metaphor of a state of nature became a reality. There was no security; kidnappings were endemic; rapes common; theft and looting rife; sectarian armies policed various areas; society fractured into tribes and families and finally to individuals, alone, terrified, defenseless. Just as peace and order, after a while, help sustain and reinforce themselves, as people adjust to the expectation of security, and so develop peaceful habits and behaviors, so the opposite dynamic is also true. As disorder persists, it also accelerates, as each person doubts the existence of government, seeks his own security first, and remains poised for combat. If the cycle is not restrained quickly, it can lead to anarchy, a final war of all against all. This was Hobbes's nightmare; and it is the nightmare that lurks much closer beneath the veneer of every civilized society than most of us want to acknowledge.

So the first goal of conservative politics is not virtue, or education, or liberty, or the integration of a divine or eternal truth into every rule and regulation. It is much more basic than that. It is security. Without such security, it is impossible to have the peace necessary to cultivate virtue, apart from the virtue of courage. Without security, we are all forced to bludgeon human personality into the uniform mold of physical strength or cunning. Without security, we cannot even afford the luxury of questioning whether we need security. We are too busy trying to stay alive and intact. There is a reason why the founding fathers of the United States put "life" as the first objective of their new political order. Without establishing the possibility of physical security, politics becomes impossible. In such a situation, politics can be expressed only in the idiom of war and violence or the threat of war and violence.

There are those who argue that, in fact, human beings are more peaceable and cooperative creatures than that. Without one entity having a monopoly of force, human beings would be able to get along more or less okay. Hobbes is exaggerating, they argue. Hobbes's response would simply be: how do you know that such civility is possible in advance? Our self-understanding is flawed enough. How on earth can we know what is inside anyone else's psyche, and what he might do next? How do we know that the next person we may bump into is, in fact, the civil, cooperative, rational human being we assume he is? If there is any doubt as to whether your random next encounter will be with the Dalai Lama or Charles Manson, your sanest option is to bet on Manson.

And you can see his point, from the perspective of doubt. You need to rest your politics, Hobbes insisted, on the worst possible scenario. Why? Not because all people are serial killers; not because there aren't many people far more civil than that; not because on average, we may get along fine with those we do not know; but for one reason and one reason only: Because we can't be certain of anything else. And the difference between being certain and not being certain may be the difference between life and death.

Even in a world where a sovereign does indeed rule effectively, Hobbes might argue, there is still crime. People are still often killed at random, walking their dogs, driving their cars, getting lost in the wrong neighborhood. If this is true of a successful political order, what would an unsuccessful one look like? Hobbes wrote at a time when England was itself divided by civil war, where, for a while, no single unchallenged sovereign ruled. He was clearly appalled by the consequences. He saw previously civilized people descend into brutal warriors; he saw families divide and murder each other; he saw pitched battles in the fields of a green and previously pleasant land. He saw the surface of society peeled back a little; and he saw the terror and violence that lay underneath. His response was to insist that politics be built on the lowest but most secure ground: doubt and, therefore, fear. If you did that, you would never be disappointed. In fact, your experience would generally be one of pleasant surprises.

Of course, someone might argue that if order is your fundamental priority, what is wrong with a dictator? And the response to that is a simple one: a dictator can harm you as well. The order of a dictatorial state is, in fact, not real order, not a genuine guarantee of security, because such a dictator can kill you as well—at random. He is as unknowable as the person sitting next to you on the subway because he is restrained by no laws and cannot be relied upon to act consistently. And, in fact, as Machiavelli noted, the most successful dictators often succeed precisely by acting irrationally, randomly, murdering opponents who merit their wrath alongside total innocents who could harm no one. The unpredictability is the point, as Saddam or Stalin or Castro remind us. The dictator can threaten your life as effectively as the thief down the street or the thug in your apartment building. He can even be more dangerous, since his monopoly of force and power can lead him to snoop, to buy off informants, to allow arbitrary arrests, to imprison without a fair trial and torture with impunity. This isn't a relief from insecurity but a capitulation to it.

Again the Iraq example, since it is so fresh, is worth recalling. Iraq, before the 2003 invasion, was more superficially stable than the chaos and anarchy that followed. But it was also a terrifying place. A lone dictator did indeed have a monopoly of power, but he maintained it by a regime of constant terror and fear. You never knew when a soldier could come knocking on your door at night; or when a son or daughter would be dragged off the street and tortured or murdered; or when a father or brother might be conscripted with a gun barrel in his back. Such a monopoly of force was almost as bad as complete anarchy. It was unpredictable, violent, terrifying, and psychologically stultifying. Enduring a dictatorship is life lived as trauma.

The question then becomes: how do we sustain order by granting some entity a monopoly of force and yet remain protected against that entity as well? The answer is that the monopoly of force must somehow be constrained and restricted to ensuring the security of everyone alike. This is not an easy feat, and its achievement in human history has been remarkably rare and fleeting. Most human beings have never enjoyed such a stable and free state of affairs. Hobbes didn't. Neither did Montaigne. They both lived in countries that were, at the time they wrote, torn apart by civil and religious warfare. Oakeshott, for all his bohemian tendencies, fought in uniform in a war to protect his own civilization against Nazi tyranny, and lived to see the capital city of his own land bombed into rubble. Socrates was executed for the crime of freethinking. Machiavelli lived in fear of the princes he tried to teach. These great imaginers of security and freedom knew insecurity and terror well.

This is why conservatives are often defined as being obsessed with "law and order." Some decry this priority as indicating some kind of sympathy with state violence, but it is actually

236 · ANDREW SULLIVAN

the opposite. A conservative values order not because he wants
to see other individuals hurt—but because he can see himself in
their place. He values it because it alone guarantees real peace
and thereby real protection from bullies and thugs. One big thug
that you can reliably restrain by a constitution and by laws that
apply to all alike is better than multiple little thugs terrifying
their neighborhoods with their capricious violence and lawless
prerogatives.

But why equality? Why should all human beings within a
given country or location be deemed equally worthy of security? Why couldn't we have a country where a majority ruled
by fiat, as long as they maintained order? You could even have
a minority of people who were deemed superior to the rest and
mandated to protect the others from harm. Remember Plato's
philosopher-kings and their guardians?

The objection to this is again based in doubt (as Plato
himself indicated). The kind of virtue and altruism required in
such an aristocratic class is rare. We have no reason to believe
it would be sustained by any group of humans for any length of
time. They'd be too tempted to abuse it. We know this because
we know ourselves, and know that our own human nature is too
fickle to be trusted.

In this world of unknowing, it surely behooves us to give
others the benefit of the doubt. By that I do not mean trusting
them with our physical safety. I mean simply accepting that the
other is not fully knowable, but can be seen, at best, as a mysterious version of ourselves. And if my neighbor cannot be known
for sure, if his inferiority or superiority is beyond my firm grasp,
he surely deserves no more and no less than I do as a political
being. His security is my security, so my rights are his. And so
an account of human life that rests itself in doubt is an account

that leads ineluctably to an acceptance of formal human equality in a political order.

Notice what this theory of equality isn't. First of all, it isn't a statement that God has made all human beings equal. After all, this bold statement begs an obvious and enormous question. Whose God? And how do you know what he or she or, more plausibly, it believes about human beings? As soon as a polity contains people who believe in different Gods or who hold different and mutually contradictory doctrines, or have no faith at all, "God" cannot be a basis for political agreement. In fact, the minute you invoke God as a guarantor of political equality, you immediately turn what might be political disputes about the practical here and now into epic struggles about the meaning of the universe itself. Religious politics is ideological politics but on a massive and profound scale. Bringing "God" into the argument merely runs the risk of dissolving a polity into factions that differ with each other on the most profound issues imaginable, and must insist on the inequality of various truths and therefore of various people who hold such truths. It is a spontaneously self-defeating move. It is a law that must almost immediately repeal itself.

If you rest your political system on theology, in other words, you do not solve your problem of establishing a common ground for a political order. You make it much, much worse. You sow the deepest division and the starkest polarization. Montaigne and Hobbes saw this firsthand: their own polities in England and France were rent into little pieces by religious disagreement—and they were all Christians! In France and England, Protestants fought Catholics in a maze of burning stakes and smashed statues. Later the king of England claimed a divine mandate to rule unfettered by parliamentary interference, invoking God again. The result was that a remarkably peaceful and successful country was plunged into civil war and the king was eventually executed.

In a much milder fashion, the appropriation of religious groups for the political base of the modern Republican Party immediately and progressively divided the United States into "blue" and "red" states, between "Godless" and "God-fearing" regions. When such a divided country was thrown unexpectedly into war, it could find no stable center over which to unite. Hence the acidic nature of our current cultural politics—and the poisonous divisions in a country that desperately needs to remain united.

Thanks to the decision to turn Republicanism into a primarily religious affiliation, a country at war with a lethal enemy found itself simultaneously at war with itself. An attempt to rest political equality on religious truth is simply a recipe for social disaster and political failure. It has always been so. But in modernity when many human beings are even less constrained by cultural inheritance than they once were, and less deferential to ecclesiastical or any other form of authority, and where the variety and depth of religious faiths is immensely more diverse than it ever has been in the past, such a political move is a recipe for disaster. And so it has proved.

It's important to note here that basing human equality on doubt isn't and cannot be a factual statement about the substantive equality of all people. Such an idea is itself, of course, silly. Substantive inequality in various abilities is something every sane human being has always known from her first consciousness. Some people are simply faster runners than others; some are stronger, or smarter, or kinder, or braver. Some of this inequality is manifestly innate; some is acquired or not acquired by experience or education or practice. But human inequality is simply a natural and social fact. No society has ever existed without it; and none ever will.

All that conservatism asserts is something much more modest than a claim that human beings are substantively equal. The conservative merely asserts that while we cannot prove or know the substantive equality of human beings, we do know enough to grant them *formal* equality. By formal equality, I mean simply the respect due to a fellow citizen, in the limited public world of citizenship. We may think that the woman down the street is a fool or a liar or a saint or a good and old friend. She may be, in our judgment, our inferior or superior on any number of measurements. But none of these qualities or flaws is civilly or politically relevant. None affects her status as an equal citizen in the formal sense. She still has one vote, as do you. She is governed by the same laws as you are. Fools and sages, old and young, male and female, gay and straight, beautiful and ugly, moral and unprincipled, strong and weak: as human beings, we are vastly different. But as citizens, the conservative argues, we are utterly indistinguishable from each other.

All this theory of equality depends on is the fallibility of our own knowledge; and the recognition that, since we cannot fully know another, we owe one another the benefit of the doubt as equal citizens of any political order. From doubt comes security, and from doubt comes equality. There may, of course, be occasions when the government must necessarily treat us differently, make distinctions between this group of people or that one. The most obvious distinction is between adults and children. A government is not going to recruit seven-year-olds into military service. But even then, in a broader context, the government still acknowledges the citizenship of a child; and the state exists to protect children in some instances, even from their own parents, if their physical security is at stake. There is, in other words, a presumption in the way a government interacts with its own citizens. That presumption is that it will treat each citizen absolutely alike, unless it has a very compelling interest or reason not to. And it is up to the government to prove it has a good reason to discriminate rather than up to a citizen to prove she is equal under the law.

You may notice how modest this political theory is. But its modesty is the point. It is austere and restrained. It makes no grand claims about humanity and rests itself in no eternal truth. It proffers no story about what is noble about humans; it makes no case about what justice means or should represent. It just accepts the fog of human existence and does what it can to prevent violence and tyranny. But why? As we inch toward the least we can say, we stumble across the thing we have been looking for all along. We have discovered freedom.

II

THERE ARE MANY WHO OFTEN INVOKE THE RHETORICAL bromides of "freedom" but, when pressed, acknowledge that no such thing really exists. The fundamentalist believes that humans have freedom—but only to choose the good; and he believes that a government dedicated to upholding that good, whether deduced from God or his own version of "nature," has every right, and, in fact, a duty to ensure that as many citizens as possible achieve that good. And so laws are designed to encourage virtue and discourage vice. Freedom is limited and conditional. A socialist will argue that freedom is an illusion to those people who begin with a material disadvantage, and that the state must act to remedy such a disadvantage before freedom can truly exist. The poor are not free, he argues. Those who are at the bottom of the heap of human inequality deserve substantive aid to equalize the system. And so he wants a system of redistributive justice to ensure "real" freedom.

A conservative, in contrast, will be skeptical of both arguments. He'll want to know from the fundamentalist who exactly came up with this "good." He'll ask why he should adhere to a view of virtue which is deduced from a religion he doesn't share or from a "nature" he doesn't recognize as his own. He'll ask the socialist, in turn, why he is being forced to give up his own money and property for the sake an idea of substantive equality that sounds like a surreal fantasy to him. Who guarantees either vision—that of the virtuous or that of the substantively equal? And who says what is virtue? And by whose standard do we judge substantive equality? If inequality remains after redistribution, what then? And, by the way, how do we control a state that has the power to divest me of my income and wealth and property?

The conservative will just keep asking the questions; and will refuse to give up his freedom until he gets an answer that satisfies his skeptical soul. Which is to say, he will never stop asking; and so the political project based on virtue or substantive equality can never get started. The pesky conservative stands doggedly in the way, knowing what he doesn't know, distrustful of those who claim to know better, querulous and quirky and always himself.

So what is freedom for a conservative? It rests, as Hobbes intuited, on being secure in one's own physical existence, and in accepting the fact that others exist who are just as human as we are, and, in political life, deserve equal treatment under the law. Then what?

The conservative will answer: everything. The great and

constant dream of the conservative is to be left alone by his own government and by his fellow humans, as much as is possible. Left alone, there is so much to explore, to do, to experience, to live and breathe and think. Life is such a mystery it demands to be taken as it is and at the same time explored further. It constantly throws us questions to which the answers are opaque and yet always inviting. Any impediment to exploring this life as fully and as intrepidly as we can is anathema to a conservative. It robs him of that thrill of being human, of that moment to be felt and experienced now, of feeling and thought and serendipity and contingency. These contingent, mysterious things are what make us human, and freedom is the means by which we live most fully as human beings. The intervention of a government is like that of a loud telephone ringing in the middle of an engrossing dinner conversation. It is inherently offensive. It commands our attention, when we would much rather be doing something else.

Montaigne again celebrated this sooner and perhaps more thoroughly than anyone since Augustine. That's why Oakeshott adored him so much. He saw freedom not simply as the ability to think alone in his study, but also to explore the world in its entirety, with as few limits and restraints as possible:·

> No judge has yet, thank God, spoken to me as a judge in any cause whatever, my own or another man's, criminal or civil. No prison has received me, not even for a visit. Imagination makes the sight of one, even from the outside, unpleasant to me. I am so sick for freedom, that if anyone should forbid me access to some corner of the Indies, I should live distinctly less comfortably. And as long as I find earth or air open elsewhere, I shall not lurk in any place where I have to hide.

A conservative is "sick for freedom." Notice that he doesn't actually have to travel to experience it. He merely has to know he can. Freedom is in the mind and soul first of all. Physical freedom from violence is a necessity because the physical body is the mortal armor of the mind and soul—no more or less. And so physical security is just a means to the real end: the freedom to think and pray and experiment in life. This freedom is the conviction that just as there is no thought I cannot think, so there should be no thought I cannot also express, no place I cannot go to, if I wish, no mystery that I am obliged to accept without further inquiry, no question I cannot ask.

This freedom inevitably has some limits. We may enter a country where such freedom doesn't exist; we may be unable to visit certain places we dream of; we may be financially unable to travel to a city we are curious about; or we may be too sick or old or young. There may also be circumstances in which our freedom conflicts with others—screaming fire in a movie theater, walking through another's property without her permission, and so on. But all these necessary limits merely reveal that freedom is a difficult enough achievement without our constructing a political order that subjects it to restraints before we have even started.

A human being alone in her own thought, inquiring into the meaning of her own existence: this is where we all start from, and it is where we will ultimately end up, even and perhaps especially at the hour of our death. And the goal of politics is to protect this freedom as much as possible in a world where we are compelled to live among one another. Freedom is the beginning of religious life; and it is the beginning of philosophical life. It is the ability to choose between each.

The Supreme Court affirmed just such a conservative understanding of freedom in the 1992 case of *Planned Parenthood of Southeastern Pennsylvania v. Casey.* The deciding justices wrote: "At the heart of liberty is the right to define one's own concept of existence, of meaning, of the universe, and of the mystery of human life. Beliefs about these matters could not define the attributes of personhood were they formed under compulsion of the State." Freedom, in other words, has to begin with our freedom to figure out who we are for ourselves in a world where so much is mysterious. It says everything you need to know about American conservatism's transformation that this defense of liberty would now be regarded as itself unconservative. It is, of course, the kernel of conservatism: freedom born out of the mystery of existence.

And the means of restraining government's necessary power is what we call a constitution. In some lucky countries, like Britain, a long history of slowly acquired liberty with few invasions, revolutions, or coups has led to a legal and social consensus that amounts to an unwritten constitution. In a new and still-young country like America, which was founded after so many European thinkers had pondered the questions of how security can be reconciled with freedom and faith, this inheritance could be translated into something more formal. And so we have something called a written constitution.

A constitution does something quite miraculous in human affairs, and few constitutions have been as miraculous as America's. What it does in the brutal world of competing human interests and opinions is to change the subject. Instead of focusing on what a polity is for, what meaning it is supposed to represent, which virtues it is supposed to inculcate, a constitution restricts itself to pure procedure. It doesn't tell us what purpose to give our own country or what purpose to give ourselves. It merely

says what the state cannot do, and leaves the rest to us. It is a su-
premely negative piece of positive action. Instead of instructing
us what we should do with our lives, it restricts itself to telling
us *how* we do it.

And so conservatism is primarily a politics of means, not
ends. Again doubt is central to this. Human beings have very
rarely agreed about what the purpose of life is, or even what
the purpose of a government or state is. But they can agree on
certain rules to be adhered to in the pursuit of whatever it is we
want to do. Such a priority on means rather than ends protects
our lives from the government and from each other. By protecting our lives and property, it protects our liberty—which begins
in existential doubt. After it has prescribed the rules by which
we shall interact, and by which the government interacts with
us and with itself, the Constitution simply says nothing. Nothing. The rest is a long, deep, fathomless, exhilarating silence.

Some religious enthusiasts insist that the American founding was a religious event; that America is a country founded on
God, and not just any God but the God of the "Judeo-Christian
tradition," whatever that oxymoronic shorthand is supposed to
mean. They usually decline to notice that the Christianity of
the founding fathers was a far more skeptical, dry version than
that of contemporary and subsequent fundamentalists. They
are aware, as Rick Santorum concedes, that the founders didn't
seem to put virtue or Christian values into the founding document of a nation. Santorum wants to argue, however, that this
is because the founders took it for granted that Americans were
Christians and would require their own government to uphold
a particular version of truth.

But the Christian context of their time makes the founding
fathers' statement even more remarkable. Yes, there are rote and

familiar invocations of the deity in the Declaration of Independence. But what's new about America—what sets it apart from the old world—is its stunning decision at the very beginning to take God out of government. The mother country, Britain, had an established church which, by the late eighteenth century, had become remarkably successful in taming religious passions and fusing the state with some religious validity. The Enlightenment thinkers who devised the American Constitution decided to do something else completely. They did not abolish God, as their French counterparts decided to do. They were too shrewd and conservative for that. Nor did they put Christian virtue and observance in the Constitution, as many of their fundamentalist contemporaries would have liked. Instead they deftly put God to one side. They insisted that government would never side with one religion over another, which means it cannot side with any religion at all.

Of course, people want to talk about ends rather than means. They want to talk about what they want to do with their lives, individually and collectively, rather than what the government should not do. And conservatism in continental Europe, less familiar with Anglo-American conservative conceptions of individual freedom and security, took this tack in the eighteenth, nineteenth, and twentieth centuries. In the Spanish-speaking world, the three pillars of faith, family, and property took precedence over individual liberty. Bismarck forged a welfare state and rallied his Protestant base with a "culture war." Franco's authoritarian conservatism brought church and state so close it was hard to tell them apart. (In some respects, Franco's Spain was a theoconservative Eden.) This form of European conservatism—which ended up in its benign forms in "Christian democracy"—was deeply suspicious of market capitalism, of what was understood as Anglo-Saxon "Manchester Liberalism." And you see the expressions of this illiberal conservatism in many of the authoritarian regimes that eventually came to prominence in South America.

In Britain Toryism also clung to God and empire and paternalism as one strand in its makeup. But British conservatism, from its inception, still treasured individual liberty. Edmund Burke, one should recall, was a Whig. He favored American independence, and questioned executive power. And British conservatism, from Burke onward, developed an approach to social change that saw the advance of liberal democracy and market capitalism as allies to be co-opted, rather than as enemies to be feared and tamed. This was Benjamin Disraeli's insight. It would be Margaret Thatcher's as well. Her conservatism was, in some respects, the final fusion of American and English conservative thought: seeing individual liberty as itself a patriotic achievement, viewing the common heritage as a product of individual freedom, which, in turn, furthered collective pride and unity. Britain was great, in other words, simply because Britons were free. Individual liberty did not detract from a collective purpose, as Continental conservatives feared. For English conservatives, individual liberty was the paradoxical definition of their collective purpose.

What America's founders realized, however, was that they had one extraordinary chance to pivot an entire nation toward these principles of freedom and restraint. The Constitution was their answer. If English conservatism had to somehow marry individual freedom and market capitalism to a monarchy and established church, Americans had no such burdens. Conservatism in America could point to the Constitution as the source of national identity and meaning; and the Constitution in turn could describe that meaning as a system in which the means always took precedence over the ends.

That's why those people who get irritated by such cumbersome procedures as parliamentary maneuvers, or rules about rules, or necessary legal hurdles to overcome before action is taken, are not really conservatives in the Anglo-American sense. President George W. Bush summed up the anticonservative position when he told the author Fred Barnes that his interest "is not the means, it is the results." The impatience of President Bush with procedures and the rule of law is integral to the anticonservative mind-set. And so he ignores international and domestic law in order to torture terror suspects to get intelligence he deems essential. And he declares he will not abide by laws that have been passed by the formal constitutional process by issuing extra-constitutional "signing statements." He is suspicious of the idea that states can decide for themselves what policies they will pursue, on marriage or medical marijuana or abortion. And if the Constitution gets in his way—as it does, for example, with abortion and marriage—he proposes amending the Constitution to achieve the ends he seeks. He is a man of results. And he is not therefore someone who understands or has ever understood what is so special about the Anglo-American conservative tradition.

The founders, of course, anticipated that individuals like George W. Bush would seek glory and honor by winning over the executive office. They knew of Aristotle's description of "thymos," the urge to gain a public reputation by doing certain great feats and directing communities and countries to great communal ends. They knew that elected congressmen and senators would seek to achieve direct aims and bend government to certain substantive ends. And they saw that even those dedicated to restraining government in the judiciary might also be tempted to use their own office to enforce certain social ends and goals. And so they set up a scheme in which much of the energy of these entities would be directed toward fighting each other, rather than achieving their own goals. They divided government so that it would be required to fight itself rather than actually getting anything done.

When people bemoan the inefficiency of American government, they show how little they understand the genius of the founders. The genius is precisely the inefficiency. By putting a constitution in place that eschews any single human good, by enshrining it in institutions that, given human nature, would spend the bulk of their energies wrestling each other into paralysis, the founders maximized the real goal of their experiment: individual freedom. They wanted government to be effective enough to maintain order and protect liberty and property, but they didn't want it to be very effective in any other respect. They created a founding document that would place a long list of restraints on government power, then they divided the weak government they had left so that it would be focused on itself rather than the people. Bingo! They constructed a government that would, in important respects, be self-limiting.

To protect against a democratic majority reshaping the country to its own idea of what the good is, they instituted such rules as freedom of speech and religion, and even ensured that individuals could arm themselves if necessary to fight back against a government that got too bossy. They also kept a vast amount of power in the individual states so that the federal government couldn't impose a national consensus on a small or quirky region. They protected the interests of sparsely populated rural regions by giving every state two senators, regardless of its size or population. They often preferred indirect elections to direct ones, and mixed aristocratic elements (Senate and Supreme Court) into the democratic mix (House of Representatives, President). And to ensure that their masterpiece would not be desecrated too easily, they made amending the constitution itself extremely hard: two-thirds majorities in House and Senate, and in three quarters of the states.

This emphasis on means rather than ends was also a brilliant way of insuring against error. A conservative is defined primarily by his profound grasp of the limits of human understanding. He knows we are always screwing things up; he knows that an idea that seems inspired one day may seem like the dumbest thing on earth a year later. He knows that humans are collectively subject to bouts of hysteria, zeal, panic, or even waves of well-intentioned moralizing. The point of the American constitution was not to banish such human fallibility, but to bank on it. And so if the federal government had some apparently bright—but ultimately flawed—idea, it couldn't impose it at once on everyone without a long and cantankerous battle. In that battle, individual states could go their own way. Or the Supreme Court could block the proposal. Or the Senate, more detached from popular passions than the House, could stop the idea in its tracks. Or if the project percolated up from the House and Senate, a president could veto it.

With any luck, most ideas for "improving" humankind, "solving" pressing problems, "ending" inequality, ridding the world of tyranny, "protecting the children" and other well-meaning abstractions would end up stymied in an interminable process of deliberation. The point was to stop anything collective from being done too easily. And the premise of this intended frustration was the founders' grasp on the limits of human wisdom. In that critical sense, they were conservatives. And in that sense, America is a conservative nation.

The great irony, of course, is that a country with a government dedicated to doing as little as possible became the most prosperous, powerful, and culturally dynamic place on earth. Less turned out to be much, much more. Leave people alone, and you'll be amazed at what they can do. And their achievements over the centuries were more durable and less repressible than those of any other nation—because they were less likely to be dragooned into some collective fantasy that would end up in tears. Not for America the fatal temptation of communism—a utopian dream that, in Russia, ended in poverty, misery, and tyranny. Not for America the unwieldy British welfare state that ended up impoverishing a once-proud nation and strangling its resources of enterprise and freedom. Not for America the poison of fascism, with its preposterous claims to advance human history and racial purity, ending in the rubble of Europe in 1945.

All these views and tendencies could be found in America, of course. There were American communists and fascists and socialists and theocrats and (especially) racists, all eager to turn this continent into their own utopia. But they kept getting foiled by the institutions and rules and procedures the founders had left behind. Every now and again, someone would do his best to fix the system, as FDR did. And it remains true that America was forever a place with less economic liberty after FDR than before him. But even he, with the collective propulsion of world war at his back, and consecutive terms in office, couldn't turn America into a centralized European welfare state. It was too hard. Resistance was always popping up in some quarter or other: court or Congress, press or pulpit. Religious fervor also overcame the American experiment from time to time. The religious right even managed one of the great dreams of theocrats everywhere: they criminalized pleasure—their ultimate dream—by banning alcohol. They were even able to amend the Constitution to do it. Somehow, however, sanity eventually prevailed, and prohibition collapsed. Liberty endured.

It transpired, as Ronald Reagan deftly understood, that government was not the solution to the human problem. Government was itself part of the problem. The solution was constitutionalism, and the transformative power of human freedom it unleashed.

III

But did America then have no ultimate point? Was it all means and no ends? Again, the founders' response to this question was little short of genius. When obliged to come up with a phrase that would encapsulate the point of their negative, do-as-little-as-possible conservative enterprise, they came up with four words: "the pursuit of happiness."

It's a small phrase when you think about it. It's somewhat overshadowed in the Declaration of Independence by the weightier notions of "life" and "liberty." In today's mass culture, it even comes close to being banal. Who, after all, doesn't want to pursue happiness? But in its own day, the statement was perhaps the most radical political statement ever delivered. And when we try and fathom why it is that the United States still elicits such extreme hatred in some parts of the world, this phrase is as good a place to start as any.

Take the first part: pursuit. What America is based on is not the achievement of some goal, the capture of some trophy, or the triumph of success. It's about the *process* of seeking something. It's about incompletion, dissatisfaction, striving, imperfection. In the late eighteenth century, this was a statement in itself. In the Europe of the preceding centuries, armies had gone to war and countries were torn apart because imperfection wasn't enough.

Enter the Americans. Suddenly the eternal, stable order of divine right and church authority was replaced by something far more elusive, difficult, even intangible. Out of stability came the idea of pursuit. To an older way of thinking, the very idea is heretical. The pursuit of what? Where? By whom? Who authorized this? By whose permission are you off on some crazy venture of your own? Think of how contemporary Islamic fundamentalists must think of this. For them, the spiritual and intellectual life is not about pursuit; it's about submission. It's not about inquiry into the unknown. It's about struggle to implement the will of Allah. Since the result of this struggle is literally the difference between heaven and hell, there can be no doubt about what its content is, or the duty of everyone to engage in it. And since doubt can lead to error, and error can lead to damnation, it is also important that everyone within the community adhere to the same struggle—and extend the struggle in a fight against unbelievers.

Today many of us find this religious extremism alien. But it was not alien to the American founders. The European Christians of the sixteenth and seventeenth centuries were not so different in their obsessiveness and intolerance from many Islamic and Christian fundamentalists today. And against that fundamentalist requirement for uniformity, the founders of a completely new society countered with the notion of a random, chaotic, cacophonous pursuit of any number of different goals. No political authority would be able to lay down for all citizens what was necessary for salvation, or even for a good life. Citizens would have to figure out the meaning of their own lives, and search for that meaning until the day they died. There would be no certainty, no surety even of a destination. Pursuit was everything. And pursuit was understood as something close to adventure.

And then comes the even more radical part. The point of this pursuit was happiness—and "happiness" in a new and radical sense of the word. Before the American experiment, governments and states and most philosophers viewed happiness as incidental to something else. For Christians, happiness was achieved only if you were truly virtuous. Happiness was the spiritual calm that followed an act of charity; the satisfied exhaustion after a day caring for others. For Aristotle happiness was simply impossible without virtue. Happiness was an incidental experience while pursuing what was good and true. The idea of pursuing happiness for its own sake would have struck Aristotle as simple hedonism.

The happiness someone feels drinking a cold beer on a hot day or jumping out of a helicopter in a parachute for its own sake was not a happiness Aristotle recognized. The happiness of Oakeshott's fly fisherman was opaque to him. The same could be said for the ski bum who spends his days climbing and descending the same mountain and feels alive only as he sails through the whiteness. The same goes for the Hawaiian surfer, the New York raver, the Ann Arbor stoner, the Alabaman hunter, and any true fan of baseball or cricket. These are activities that were once categorized as mere hedonism or vice or inanity. That they might all be ways in which human beings pursue their own happiness and learn their own ways of figuring out the world would not have meant much to many human beings in the distant past.

They had far more pressing things on their mind. Which is to say, they never knew such freedom.

For almost every pre-American society, other goals clearly had precedence over this dangerously subjective pursuit of well-being. Remember Cromwell's England? Or Robespierre's France? They weren't exactly happy-fests. Cromwell even banned Christmas! Again, in radical Islam today, American notions of happiness—choice, indulgence, whimsy, humor, leisure, art—always have to be subjected to moral inspection. Do these activities conform to religious law? Do they encourage or discourage virtuous behavior, without which happiness is impossible and meaningless? These are the questions human beings have always historically asked of the phenomenon we call happiness.

Not so in America. In America happiness is an end in itself. Its content is up to each of us. Some may believe, as American Muslims or Christians do, that happiness is still indeed only possible when allied to virtue. But just as important, others may not. And the important thing is that the government of the United States, if it is true to the spirit of its founding, takes no profound interest in this argument. All that matters is that no one is coerced into a form of happiness he hasn't chosen for himself—by others or by the state.

Think of this for a moment. What America means is that no one can forcibly impose a form of happiness on anyone else—even if it means that some people are going to hell. Yes, there have been many exceptions to this over the years—and America has often seen religious revivals, spasms of cultural puritanism, cultural censorship, McCarthyism, the vilest racism. But the government has been barred from the deepest form of censorship—the appropriation of any single religion under the auspices of the state. You can call this all sorts of things. In my book, it's as good a definition of freedom as any. But to others—countless others—it seems a callous indifference to the fate of others' souls, even blasphemy and degeneracy.

Most other countries have acquired identity and culture through ancient inheritance, tribal loyalty, or religious homogeneity. Even a country very like the United States, Britain, still has a monarchy and an established church. Other developed countries, like Germany, have succumbed at times to the notion of race as a purifying and unifying element. Many others, like Pakistan or Saudi Arabia, cling to a common religious identity to generate a modicum of political unity. In none of these countries is "happiness" even a political concept. And in none of these places is the pursuit of something in and of itself an admirable goal, let alone at the center of the meaning of the state and constitution. One of the first things the Taliban did in Afghanistan was to forbid the flying of kites. After all, what's the point of flying a kite? What the Taliban did not get—what they, in fact, actively opposed—was what free people feel in their bones: the pursuit of seemingly pointless happiness is the point.

Let me put this another, perhaps more shocking, way. The freest society is one in which the quintessential, ultimate activity is play. Security is guaranteed; work is done; the wealth that freedom creates enables leisure; and leisure begets play. When we play games, we suspend for a time the burdens of practical life—of earning a living, feeding our bodies, getting enough sleep, saving our souls. We engage in activity that has no point; and those who play games merely to win them miss the point of playing. Games help train us in the restraint, prudence, and cooperation that are central to democratic life. They teach us activities that lead nowhere but where they are.

Games, mind you, are very dependent on firm rules. In fact,

without rules, games cannot exist or degenerate very quickly into fights. The best games are those which require no umpire or referee. The players themselves are mature enough to put aside their own egos and engage in the competition for its own sake. This is what we call sportsmanship. But sometimes umpires are indeed necessary.

Think of the constitution as the ultimate umpire in the game of all games called politics. Yes, we can get extremely passionate; in very tough games, tempers may even flare. But in this game, there is no final score and no ultimate winner, because there is no collective goal in America except the freedom all citizens retain. And so the wise statesman keeps his cool and maintains his reserve and plays the game with verve and imagination but always with the knowledge that any resolution will be temporary and cede to a new problem, a new predicament. We play because we have to live with one another somehow; and we have to resolve the conflicts and differences that are inevitable in any human society and that are constantly changing shape and dimension. But the only thing that matters in the end is that the rules are observed.

Yes, small victories can be won, and one party or other will achieve its goal or enact its program. But these victories are inevitably followed by defeats. In fact, every political career ends in failure. It must—because every substantive human endeavor ends ultimately in failure. Things change and what you were trying to solve disappears and a new problem takes its place; or your solution itself creates another set of problems and they require another fix, which leads to more complications. Even a real win—such as a political realignment, the holy grail of partisan machinators—always carries within itself the seeds of its own self-destruction.

And so a president reelected to a second term subsequently sees his poll numbers collapse; and a president who retires to accolades is subsequently forgotten by history or, worse, is later reinterpreted as a crashing failure. Reputations crest and crash through the ages. Political battles that seem epochal at the time eventually succumb to the faded indifference of later generations. Every now and again a figure emerges who rescues a country from calamity—a Queen Elizabeth I or an Abraham Lincoln. But these men and women are rare; and in successful polities, with wise foundings or lucky geography, they are mercifully unnecessary.

IV

THE KIND OF POLITICS—AND IDEA OF POLITICS—THAT I have just described is unnerving to some people. They argue that it is philosophically hard to defend. They say that a politics based on doubt and contingency is vulnerable to an argument based on something more positive and substantive. And so some have argued that the West needs a revived notion of Christianity to defend itself against the reemerging force of theocratic Islam. Or they insist that we need to recover the argument that nature itself has granted all human beings inalienable rights to freedom. Several conservative thinkers, dismayed by what they see as the decadence of Western culture and society, insist that we too can have the certainties of other less world-weary civilizations, that we can base freedom and constitutionalism on something philosophically clear and certain and convincing to the impartial observer.

But these alternatives tend not to turn out so well. The argument for a certain kind of politics based on "natural law" begs as many questions about what nature is as it answers. The argument for our freedom based on "natural right" is similarly suspect. The case that "nature" proves the inalienability of human freedom may have its emotionally and rhetorically powerful moments, but it is itself a philosophical argument, resting on a philosophical distinction between nature and custom, and so is subject to philosophical inquiry. Once it is itself recognized as merely a postulate to be examined, it becomes far harder for it to work as an easy and secure public rampart for Western freedom. It gets caught in the skeptical web.

The most famous of conservative thinkers behind the attempted resuscitation of such a "natural right" was Leo Strauss. Strauss, a refugee from Nazi Germany who taught for many years at the University of Chicago, is the object of much fevered speculation among those who have never read him or have read him briefly or read only those who claim to have read him. But when you actually do read his own carefully chosen words at length, you encounter a mind far removed from the glib certainties of today's religious fundamentalists, or even some of his own "neoconservative" disciples. You find instead a deeply serious mind inquiring relentlessly into the eternal questions that bedeviled Hobbes and Aristotle, Montaigne, and Oakeshott.

In preparing for this book, I spent several months reading Strauss as carefully as I could. I had been taught by people who had been taught in turn by Strauss, and I had already read some of his writing, so I had some provisional idea of what he was about. But I wanted to make sure I hadn't missed anything that might give credence to his more paranoid critics and more zealous acolytes. But, as I suspected, he is nothing like the caricature drawn by polemicists on both right and left. He was fundamentally a student of texts, and a quirky, often funny, always penetrating interpreter of various great works of political philosophy.

What I learned from his writing and from those who learned from him in person was indeed the kind of skepticism and So- cratic reasoning that is at the heart of the conservative tempera- ment. His great contribution was to impart to his students a sense that the great questions of human existence—reason or faith, Athens or Jerusalem, democracy or aristocracy, virtue or freedom—were still open. Nothing was settled, except the per- manence of the problems and the alternative solutions. Every- thing was worth questioning—now as much as ever before in human history. The blandishments of historicism, and the anes- thetizing assurance that philosophy itself had become no more than language games and logic puzzles, were arguments worth resisting, he argued. There was a world of ideas to reconsider afresh in every generation.

But he added something else: a very conservative sense of loss. Modernity, he sensed, had lost something that may have been available to earlier less disenchanted times and places. The effortless unity of ultimate meaning and political life—torn asunder by modern doubt and freedom—is, indeed, perhaps worth longing for. The ancient virtues of courage and wisdom, temperance and friendship have perhaps been eclipsed by the low and basic ramparts of the modern state, dedicated to secu- rity rather than nobility. Strauss worried, for good measure, that democracy might be too soft to stand up to the brutal dictator- ships of the twentieth century; and some of his followers still worry that the West no longer has the conviction to stand up and defend itself against the resurgent certitudes of Islam.

And so it may be useful, some of his followers argued, to publicly support those who retain core elements of unquestioned biblical faith, in order to have some people who could defend a country and a civilization vulnerable to barbarians. Strauss worried that doubters make poor soldiers. And he worried that Socratic skeptics, especially those who had read and thought they understood Nietzsche, might teeter over into nihilism and relativism. He feared that the substantive achievements of despotic regimes, like that of Soviet Russia or the Nazi Germany he fled, might topple the sometimes easygoing, comfortable, and internally querulous West. He saw appeasement in the 1930s, as many did, as proof of democracy's vulnerability, of the hollowed, empty center of an unillusioned West. Without Churchill, might we not have lost everything?

This is an understandable and profoundly conservative worry. But there are two things to be said about it. The first is that Strauss himself was no fundamentalist Christian or Jew. He was a philosopher. And he knew that philosophy's chief rival was revelation. He respected revelation deeply, but he was a doubter himself. His reading of many thinkers strips them of their Christian surface to reveal the atheist beneath. After you have read Strauss on Locke or Hobbes, it is hard to read them the same way ever again. And after you have read him on Machiavelli, it is hard to stop laughing on every page.

So any attempt to infer from Strauss an esoteric plot to prop up religious elements in politics is to impute to him a kind of cynicism his writing does not merit. His point was not dishonesty, it was reason; it was the pursuit of constantly elusive truth, even if that truth may not be easily available to those who aren't sufficiently intelligent or intrepid. He didn't run for office, he conducted seminars. He was not a political figure, although he was fascinated by the relationship between politics and philosophy, between the city and the thinker.

His deepest teaching, it seems to me, was about the precariousness of the philosophical calling, since a philosopher must necessarily live in a political world, and yet his constant questioning will inevitably unsettle majority prejudices and customs. And so, just like Oakeshott, he insisted on a separation of theory and practice, of politics and thought. And just like Oakeshott, he was determined to protect the university as a seat of truly unfettered liberal learning, as the guardian of the intellectual freedom which is the lifeblood of the Western experiment in free humanity.

He also clearly believed that whatever the merits of an aristocratic order, it was no longer possible. The authority of priestly or political elites had long since disappeared in the dizzying, changing, rebellious spirit of modernity. None of this could be put back together. And while there were many things to regret about what was lost, there was also a great deal to be said for the democratic life that remained. Above all, it was the political system that best allowed a philosopher to think freely. And that was Strauss's ultimate goal—to secure a space, however small and fragile, for an individual to ask all the important questions, and to read and ponder all the great texts that had explored them, free from censure or pressure or threat. He would have had a great conversation with Montaigne. Perhaps, somewhere, that conversation is still going on.

If Strauss's followers somehow believe that he would have been unperturbed by the rise of explicitly sectarian political movements, by the fusion of theology and politics that has come to define contemporary American conservatism, then they should read him again. For Strauss, freedom was a profound value. It was not without cost—but then what is? He defended intellectual freedom against the intolerance of the New

Left in his time, but he would defend it against other enemies in other places and eras. His spirit is kindred to Oakeshott's. Both leavened their deadly seriousness with humor and mischief and love of the philosophical chase.

History, moreover, has proven Strauss's worries about the tenacity of the West exaggerated. It turned out that the most democratic country on the planet became the wealthiest and strongest. The human energies that constitutional democracy unleashes generate wealth and power and strength that far surpass the rigid orthodoxies of theocracies or socialist utopias. In the great battle of the cold war, which he worried the West could not win, the outcome, in the end, wasn't even close. The West achieved not just victory but utter vindication. Even in its spasms of self-criticism and self-hatred, it was more resilient than communism.

The conservative fear of modernity, understandable in the beginning and middle of the last century, should surely be abated now. Even an America saddled with an unwieldy welfare state, stagflation, a divisive war, and cultural and social decay in the 1960s and 1970s recovered to triumph by the 1990s. The resources it found to regenerate its energy and conviction came from within. A leader, Ronald Reagan, emerged to articulate them; another remarkable figure, Margaret Thatcher, gave freedom a new lease of life in Europe. Those countries that adopted individual liberty rather than collective goals as their guiding principle prospered and thrived, while others collapsed or became parasitic on Western ideas and innovations.

Cultural conservatives who predicted social decline were also proven wrong. The consequence of the 1970s was certainly what one might expect from a well-meant attempt at social engineering. But human beings recovered from the bid by government to make their lives better. A Democratic president reformed welfare and balanced the budget and harvested a peace dividend. In the silver age of limited government, the 1990s, social indicators improved markedly: crime collapsed, abortion rates declined, minorities slowly integrated themselves into the mainstream cultural fabric. And when the next threat emerged—in the form of Islamic ressentiment—Americans immediately united in defense of their way of life. The spontaneous outburst of patriotism and resilience after 9/11 showed that a secular, liberal democracy knew what it stood for and what it stood against. It took the toxic combination of religious fundamentalism, leftist anger, and political incompetence to tear the country apart over Iraq.

That this society no longer represents a philosophically unified and substantive whole is a loss greatly outweighed by the exuberance and genius and creativity that freedom has unleashed. Miracles in science and technology, astonishing advances in communication, the empowerment of millions to experience freedom of thought independently of big corporations, governments, or expensive printing presses: these achievements of free people expanded the possibilities of human freedom still further. The attack on the West by Islamism was not a function of the West's weakness, but a nihilistic, embittered swipe at a success that cast the dreary failure of so much of the Muslim Middle East into a shaming shade. It turned out our flaw was not our softness, but our strength.

When asked to defend the contingent and foundationless conservatism I have sketched here, this should be enough. We like it here. We love our way of life. The proof is in the millions who long to be here, who aspire to this dream of human potential, who yearn to escape the stifling constraints of oppressive government interference or brutal theocratic tyranny. What greater argument need we have? Our only weakness is self-doubt, which is part of our own querulous, paradoxical strength. The achievement of this freedom is a consequence of luck and tradition, history and thought, of great leaders in dark times and ordinary people in the unlikeliest of places. But it is an achievement nonetheless. We can touch it with our hands and express it with our voices. It is more secure than any abstract argument or esoteric thesis. It is as good a philosophical defense as we shall ever have. Why on earth should we ask for more?

V

OF COURSE, A CONSERVATIVE POLITICS WITHOUT ANY firmer philosophical foundation than doubt and contingent historical inheritance begs a question. Once a constitution has been set up and abided by, by what principles do we actually govern ourselves and tend to the inevitable conflicts and problems that emerge from an always changing society? Don't we need an ideology to help us settle these disputes—an ideology that rests on truths freely asserted and aggressively defended? On what grounds would we change a law or enact a government program if we have nothing but doubt to go by? Would we not be paralyzed by indecision?

The first thing to say is that, yes, indeed, the skeptical conservative temperament will often lead to paralysis, which to a conservative is often preferable to change. And in general, a conservative will want the government to do as little as possible. And where a society has already acquired the strangling architecture of social engineering, the institutions set up by a benevolent few to better the existences of the less fortunate many, a conservative may have nothing better to do than get rid of them. The great achievements of Reagan and Thatcher were negative ones: Reagan, with Paul Volcker and then Alan Greenspan, ended the Keynesian nightmare of stagflation by insisting on monetary discipline, regardless of its short-term impact. He cut a swathe through an immensely complex tax system, simplified and reduced it. Thatcher undid the inheritance of union power merely by repealing laws that protected their bullying and intimidation. She began the process of Britain's economic renaissance by abolishing currency controls and unraveling the "progressive" tilt of the taxation system, so people could actually keep much more of the money they earned, and spend it as they wished. The great conservative achievement of the Clinton-Gingrich era in the 1990s was boring old fiscal discipline and welfare reform, which abolished an existing program.

In the modern world, in other words, conservatism often means repealing laws, abolishing unnecessary institutions, getting rid of needless government departments in order to let people make their own choices as much as possible. A free market is critical to this, but not because it somehow succeeds in creating wealth. A conservative is uninterested in wealth creation as such, although he will regard the individual acquisition of property as a helpful ballast to responsible citizenship. For a conservative, it is not a criterion for a successful country that it increase its GDP by a certain amount each year. Freedom is what matters—and that includes the freedom to be inefficient and indolent if that's what someone chooses and can afford.

An economic conservative is merely interested in locating the decisions to buy this and not that, to spend here and not there, to trade this rather than that, as close to those on the ground as possible. Why? Because individual actors in specific circumstances are better judges of what they want than some distant bureaucrat or enlightened theorist. Why? Because human judgment is fallible and the closer you are to a choice, the likelier you are to make the right one.

Again the argument for free markets is rooted in doubt—not in ideology. Our ability to know things for sure in a contingent and mysterious and dynamic world is always flawed. But it becomes progressively more so the more distant we are from an actual decision in a real time and specific place. The market is a better mechanism for sorting through the immeasurably complex sum of human needs and wants than any single rational actor. It is less prone to make mistakes; and when it does make mistakes, it is more likely to correct them quickly. The government does play an important role here, of course. It must act aggressively to ensure that markets are truly free and not subject to cartels or distortions; it must ensure that money is sound and that inflation is low.

But all its actions in this respect are that of an umpire of a game not of the director of a movie. A director is supposed to have imagination, vision, the prerogative of intervention in matters large and small. It is his movie—even if he is ultimately answerable to his investors and audiences. But it is not the umpire's game; he is merely required to ensure there are no fouls, and to mediate occasional disputes. His temperament should be steady, dull, imperturbable, dry, dispassionate, disinterested. A conservative likes his life vivid and his politics dreary. In fact, he is very often conservative in his politics so he can be securely radical in his own private life. When you're off on a private adventure, you want to make sure there's a grown-up minding the public store.

How does a conservative address the question of political change which is not merely the pruning and culling of various attempts to dragoon people into successive pseudo-utopias? The answer given by Oakeshott is summed up in a telling phrase. A conservative doesn't consult his book of ideology to see how to proceed. He "pursues the intimations" already implicit in an emerging society, and adjusts law and regulations to accommodate the new realities.

One famous example of just such a pursuit of intimations was the Tory prime minister Benjamin Disraeli's decision to back universal suffrage in the nineteenth century. "Toryism" if turned into an ideology would have rejected this as a negation of its own identity and meaning. Tories were defined by their adherence to the prerogatives of the monarchy, nobility, and rural landed gentry. The idea of bringing vast masses of untutored and possibly radical working class voters into the political system seemed like socialist revolution. Disraeli differed. He saw that England was changing, that the industrial revolution was urbanizing Britain at a rapid pace, that the masses were acquiring economic power and leverage, that they were susceptible to being co-opted by dangerous and radical forces. He intuited that the job of a conservative was to deal with changing social reality. So he proposed co-opting the working classes for Toryism, giving them the vote, appealing to their patriotism, and remaking conservatism in his time.

Tradition is not a static entity. Although conservatism leans toward regretting change and loss, it is not wedded to the past. It never seeks to return to a golden age or a distant past: when you hear that voice, it is a reactionary and not a conservative speaking. Conservatism's subtle grasp of historical contingency can even allow individuals to make occasional bold moves to integrate what is new into what came before. And so Lincoln, to take another conservative example, shifted from a limited defense of the Union into a bolder emancipation of slaves. He pursued the intimations of America's convulsive civil war into something bolder and more permanent—a healing of the great wound of slavery in America's founding. Thatcher saw the untapped aspirations of Britain's working classes and extended home ownership to them, radically liberating them from the state's control. These were policy decisions designed not from some external source, but from the internal dynamics of the evolving societies these statesmen and women governed.

To take a contemporary example, the issue of rights for homosexual couples has come to the fore in most Western countries these past few years. Why? Some would argue that a handful of radical activists foisted a theory onto an unwilling public. The facts suggest otherwise. What they show is that in the 1970s, 1980s, and 1990s, social and economic change led to more and more homosexuals being open about their sexual orientation, being much more visible in their own families, workplaces, and TV shows, and less afraid and ashamed. The scourge of AIDS and a boom in lesbian motherhood also brought the legal fragility of gay relationships into the foreground. A conservative in government expects such changes in society as time goes by. His job is to accommodate them to existing institutions. He might come up with a solution like civil unions; or, worried that setting up a less demanding institution might undermine marriage, he might argue for co-opting gay couples into the existing social institution in one fell swoop. He might think it's wise to try this out in a few states first. But he will understand that some adjustment is necessary because the world changes, and the job of the conservative is to adjust to such changes as soberly and prudently as possible.

270 · ANDREW SULLIVAN

Notice what this isn't. It's not a declaration about the ultimate morality or otherwise of marriages for gay couples. That
is left to the churches or synagogues or mosques or university
seminars. It's not an assertion that gay couples have a God-given
or naturally required "right" to marry, as some liberals might argue. It's simply tending to a felt social need by an imaginative
political adaptation. It is a conservative move. A radical may
want to abolish or privatize civil marriage. A fundamentalist
will assert, as President Bush did, that civil marriage is a "sacred"
institution, ordained by God, and that the civil laws, regardless
of social reality, must conform to biblical revelation. A conservative will escape both traps.

This kind of pragmatism rests itself neither in some truth
nor in some abstract idea of nature or revelation. It takes a society as it is and adjusts it from time to time as circumstances
seem to require. Sometimes those circumstances require a radical
undoing of corruption or decay—the task of Reagan, Thatcher,
and Lincoln. Sometimes it will require a modest and restrained
form of governance that seeks merely to keep the show on
the road and government as harmless as possible. You can see
this variation on the conservative theme in the presidencies of
Dwight Eisenhower and George H. W. Bush, two of the most
quietly effective presidents of the last century—or the equally
underrated premiership of John Major, an ordinary bloke who
legitimized and entrenched the Thatcher reforms and made a
Blairite opposition party possible.

And this is conservatism's great philosophical advantage over
liberalism. It can be more flexible. A theoconservative might argue that the America I have described—of free individuals on a
constitutional order with limited government—is a product of liberalism, not conservatism. And they would not be entirely wrong

in an abstract sense. The deep belief that people have today in the Anglo-American world that they have a right to free speech, to their own property, to their own choice of faith, is indeed based on many classically liberal ideas. But over the centuries this has created an actual, empirical, unavoidably diverse society that conservatives have no choice but to govern as it is. Turning back the clock to an era of greater deference or aristocratic order or religious authority is an impossibility for conservatives who grasp the complexity of human society. And so contemporary conservatives accept this changed world and accommodate themselves to it. Maybe they will try and restrain some of its worst impulses, or seize on new opportunities for growth and development. But they will start from where they are. Because there is no other place for a conservative to start from.

And where we start from will always be changing. Each moment in history will suggest the pursuit of different intimations within any given society. That's why I cannot and will not propose a policy prescription for the present day—because there are several plausible alternatives and it is up to political actors, not writers, to intuit and develop them. But there are a few rough and ready dispositions in domestic and foreign policy that are recognizably conservative, in the sense I have tried to convey.

Domestically a conservative will seek to ensure that the freedoms enshrined in a written or unwritten constitution are protected. That is his first task. If a government starts to attack individual liberty, invade personal privacy, increase his taxes, or burden him with regulations, a conservative will resist. Aware that government can easily metastasize and that freedom can be lost in well-meaning increments, he will insist that the government remain solvent as far as is possible. He will want to keep budgets balanced so that the hard empirical choices all societies make will not be obscured or put off to the next generation. He will seek to keep taxes low and simple. He knows that a complex tax code benefits the powerful and wealthy and attracts lobbyists and parasites like flies to a picnic. Simplicity and transparency matter. He may switch his vote from party to party to ensure that government is as deadlocked as possible and capable of doing as little harm as practically feasible.

He will be particularly alert to moments when governments remove liberties for the sake of something called "security," or when the executive branch declares itself above the law, free to ignore the law, detain people without trial, or qualify habeas corpus. There may be occasions in wartime when liberties may be abridged—but only for a limited period of time, and only with insistent oversight. A president who reserves the right to wiretap citizens without a warrant, or to torture already-detained individuals is perhaps the nemesis of a conservative. If the root of conservatism is a defense of the physical security of the self from others and the government, then torture is not some unfortunate by-product of a rough and ready war. It is an attack on the constitution itself, on liberty at its core.

A conservative will also understand that one of the great bulwarks against abuse of government power is simplicity. The larger and more complex a government gets, the harder it is for most people, busy about their own lives, to keep an eye on it. So a conservative will favor as simple a tax code as possible, and will be suspicious of any government's decision to tax some people at different rates than others. Such "progressive" taxation is better described as unequal taxation, the treatment of some citizens as inferior to others and therefore subject to higher levels of taxation. The fact that their inferiority is related to their success or enterprise makes the whole enterprise look remarkably fishy. And fishy it is: it is rooted in an attempt at social engineering, to redirect the property of the successful to the less fortunate by force.

This is not to say that a conservative government will be unaware that the vicissitudes of nature, luck, and progeny will not require some efforts at amelioration. All governments have faced problems of poverty, disease, lack of education, irresponsible or nonexistent parenting. There will always be individuals whose plight is not of their own making; or some whose folly has led them into difficulty or despair that may, in the last resort, require government to step reluctantly in to help. No government can prevent such misfortune; no government ever has. But a government can do substantive, practical things to help ameliorate the worst of it.

This is the difference between a conservative and a libertarian. A conservative believes in a small but strong government. He doesn't want to abolish government because he believes that only government can secure freedom. In fact, he may well aspire to great pride in his own government and its actions. As a small government conservative, I can still take enormous pride in the fact that the U.S. military helped liberate half a continent from fascist tyranny and, half a century later, helped liberate the other half from communism. I can also feel pride when an effective government manages to provide good public education to those unable to afford a private one. I grew up in a family that could not have afforded private education. My own parents didn't attend college. It was the government that made it possible for me to get the tools necessary to experience freedom at its fullest, and I will forever be grateful.

And so a conservative government will focus on a few key tasks and do them as well as it possibly can. A conservative government should provide access to basic health care for the indigent (but no more); it can provide shelters for the homeless; it can and should offer educational opportunities to those who would not otherwise be able to afford them until the age of eighteen; it can make sure that the poorest have resources to save for their own retirement or disability. There are also public goods that any government must provide. Clean air and safe streets, secure borders, national defense, and roads, air traffic control, nature preserves, and the like. And one of its core tasks will be to govern effectively during emergencies or natural disasters. In situations like a flood, a hurricane, or an earthquake, only government can provide certain goods. It should do so efficiently and swiftly and spend a great deal of energy on ensuring that it is always prepared for calamity. Compare that with what has happened to government in America today. It is busy building pork-barrel bridges to nowhere, but stands by helplessly when its own citizens are at the mercy of natural disasters. It hopes to visit Mars, but it is incapable of securing its own borders.

In these necessary efforts in relieving the poor and disadvantaged, and in securing basic public goods, a conservative government will nevertheless make sure not to overstep its bounds. It will recognize that imparting excellence in education is extremely hard and requires local control, accountability, and the ability to hire and fire teachers at will and reward them on the basis of their performance. It will recognize that a truly effective and civil police force is an immensely difficult task and requires constant, vigilant attention. It will recognize that when people are led to believe they do not need to take full responsibility for their own lives, when they are reassured that someone else will always pick up the tab for their own decisions, they may too easily be subject to dependency. This is just human nature. We are all equally susceptible to it. Collectively we learned this the hard way: the great war on poverty of the 1960s and 1970s actually helped entrench it, precipitating and accelerating the poor family's descent into near-collapse, and thereby making an escape from poverty all the harder.

Social policies that focus on discrete and recognizable public needs, that efficiently marshal limited public resources for the betterment of clear social problems, and that never tip over into the imposition of abstractions like "social justice" or equality of outcomes, or "diversity," or other such dubious enterprises: these are defensible, even admirable, from a conservative perspective. Security is obviously the priority. Conservatives will be primarily concerned that property is safe, that borders are secure, the government's basic tasks are taken care of before anything else. And in a world where international terrorism is the greatest threat, a conservative government may well decide to shed elaborate redistributive schemes—promising free prescription drugs for the elderly or retirement income for those who already enjoy affluence—in order to prioritize security. Governments that are busy subsidizing farmers, protecting manufacturers, doling out tax breaks to supporters, financing the arts, and pursuing harmless pot-smokers tend to be distracted from the core things we most need from government. And so we have a system where the government spends more and more and yet delivers less and less of what it was originally set up to provide.

In foreign affairs the conservative will be drawn in two directions. The first is a cold-blooded assumption that there is violence out there and it is the job of the state to protect its citizens. The Hobbesian conservatism I have outlined sees the international system as one without a Leviathan, where no one has a monopoly of force, and no one therefore can enforce the law. He will therefore be vigilant and well armed. He will expect the worst, and will occasionally have to use force to preempt or respond to threats. He will certainly realize that mere retrenchment into his own castle will not protect him. Such isolationism may only delay—or even precipitate—the reckoning.

At the same time a conservative will know that the world out there is a mysterious and often opaque place. Massive cultural differences may make decision-making and secure judgment very hard. If he uses force, he will try and do so with as many allies as possible, and will have meticulously prepared for every conceivable eventuality. He will always expect the worst. He will know that in wartime, even the most careful plans can be upset by a collision with reality. And so he will never rush in, and will risk as little as possible.

A conservative will also eschew any grand notions of history or great crusades. He would never state that his goal is to end tyranny on earth or other such utopian fantasies. But he will also be aware that the fragile achievement of the West in the past couple of centuries is a blessed thing, and one that may well abate some of the violent and primordial tendencies in the larger world. So he'll do what he can to nudge those freedoms along elsewhere, always with the understanding that there is nothing inevitable about their success, and always aware that other cultures and places may express freedom in ways very different from ours. The more countries that share the conservative's view of the world, that enshrine individual freedom within their own borders, the less necessary wars and conflicts may become.

And this is a very good thing. Conservatives, after all, hate war. Their domestic politics is rooted in a loathing of civil wars and violence, and they know that freedom is always the first casualty of international warfare. When countries go to war, their governments invariably get bigger and stronger, individual liberties are whittled away, and societies which once enjoyed the pluralist cacophony of freedom have to be marshaled into a single, collective note to face down an external foe. A state of permanent warfare—as George Orwell saw—is a virtual invitation for domestic tyranny. All sorts of excrescences can be justified under the guise of security and necessity. Even when wars end, collectivist habits remain. Winston Churchill realized that the huge increase in state power required by the war against Hitler would, in peacetime, be transferred to domestic ends. And so it was. Defending freedom from Hitler's evil meant conceding it to the much milder but still corrosive pretensions of socialism. It took generations to recover. Today in America amorphous wars on terror and drugs can become systems of law enforcement or "homeland security" where government power inexorably grows and individual freedom withers.

At the same time, a conservative is not a libertarian. He knows that real threats are out there: threats from state actors and, increasingly, fundamentalist terrorists armed with far more destructive technological weapons than ever known before in human history. He knows that the ultimate responsibility for those lines at the airport lies with the enemy, not with his own government. But he'll keep an eye on his own government as well.

A conservative foreign policy will not have an "ism" attached to the end of it. A conservative understands that "realism" can miss critical internal dynamics of a foreign culture and society, and so make errors in diplomacy or warfare. He'll be aware that the same can be said for "internationalism." He knows that commerce matters and that its advance often lessens the chance of warfare. But he won't fall for the idea that globalization or free trade will eventually abolish warfare; or that we are somehow progressing to an era of perpetual peace, undergirded by international trade. He'll be open to the idea that democratization of highly dysfunctional regimes in the Middle East may ultimately be the only way to defang the pressures that have led to Islamo-fascism. But he won't be surprised if the path is a tough one, or that it might lead to unintended consequences, or blowback, or even more violence in the short term. If a conservative is skeptical of "isms," he's just as leery of "isms" that have a "neo" attached to the front of them.

This may seem like a cop-out to some readers. I haven't provided a new doctrine, or a new set of policies. But the conservative insight is that every decision in diplomacy and war is sui generis; and any doctrine in such matters will fail almost as soon as it is elaborated. There are no inviolable historical laws, no "isms" to guarantee success, no ideological guidebooks to effective foreign policy. It helps to know history; but a statesman is a person in a particular place and time, always with limited knowledge, required to decide the gravest of matters under always changing circumstances. Beyond broad parameters about what he shouldn't do—like embarking on utopian adventures or acting recklessly or believing he is on a mission from God—it is impossible for a conservative to say what any given statesman should do in any specific case. It will always depend on a unique and constantly changing blend of actors and factors that only individual prudence in real time can resolve.

Lord Halifax and Neville Chamberlain, in other words, were reasonable fellows who happened to be dead wrong in the 1930s. Churchill had a career of spectacular military failure behind him before he got the one big decision right. In the Middle East wars of the last two decades, the first president Bush was prudent but not bold enough; the second was bold but not prudent enough. That's easy to say now. It wasn't so easy to know at the time. Every now and again, we are lucky to throw up leaders who seem to have an uncanny sense of what is needed at a particular moment in time: Lincoln at Gettysburg, Reagan at Reykjavik, Churchill in Fulton, Missouri. But even Lincoln, Reagan, and Churchill got things wrong. Along with practical wisdom, experience, and a deep sense of history, there is also something called luck. It's the flip side of the historical contingency that makes freedom possible.

VI

IF WHAT I HAVE OUTLINED HERE SOUNDS LIKE A DISMAL or uninspiring approach to politics, then I have succeeded in getting my point across. Politics, for a conservative, is a necessary activity, but it should never be an uplifting one. Americans in particular often balk at this. They like bold leadership, visionary rhetoric, and great challenges. But the success of America is that its constitution does not require these things in politics for the country to work or be successful. A president or senator or governor may be appreciated for his skills in the bully pulpit, but his real job is merely to enforce existing laws, fix emerging problems, and leave the sermonizing to the real pulpits and the creativity to the country's real leaders. The real leaders of a free society are not its politicians. They are its artists and laborers, scientists and teachers, bloggers and social workers, sportsmen and movie directors, day traders and research students, architects and farmers, waiters and comedians. The great strength of a free society is not its political leadership or its government, but its people and their daily encounters with one another and reality.

A free society's deeper asset is its restless, spiritual core. By this, I do not merely mean its churches and synagogues and mosques, although I certainly do include them. I mean its platoons of volunteers, its mothers and fathers, its charity workers, its poets and dreamers, its freaks and experimenters, all those who bring charity to the people they meet and imagination to the universe they inhabit. If you want a monument to a free society, don't look at the statues in its cities, or the statistics on economic growth, or the caliber of its president. Look at its people, and what they do with their freedom.

Nothing is more important in that freedom than the ability to take the mystery of life and make something of it, to explore the destiny of one's own soul and to risk it at every moment. We will often be distracted from such an endeavor, or terribly afraid of it. We will want a political leader to live our lives for us; or a book to guide us; or a doctrine to reassure us that we are already saved or can surely expect a better future. But a conservative understands that in the face of great and constant loss, in the teeth of disorienting change, there is always the challenge of not knowing for sure. The only peace we will ever really know is the peace that comes from accepting such doubt and turning it into living. We may not know for sure where we came from or where we're going; but we can defend the lucky inheritance of our freedom, be proud of it—and live it. Now—while we still can. Now—which is the only time there is. Now—with the joy that only liberty allows and only nerve secures.

Select Bibliography

The following works are either quoted in the text or helped inform and deepen its arguments:

Allen, John, Jr. *Cardinal Ratzinger: The Vatican's Enforcer of the Faith.* New York: Continuum, 2000.

Barnes, Fred. *Rebel-in-Chief: Inside the Bold and Controversial Presidency of George W. Bush.* New York: Crown Publishers, 2006.

Bawer, Bruce. *Stealing Jesus: How Fundamentalism Betrays Christianity.* New York: Crown Publishers, 1997.

Bloom, Allan. *The Closing of the American Mind.* New York: Simon and Schuster, 1988.

Botwinick, Aryeh. *Skepticism, Belief, and the Modern: Maimonides to Nietzsche.* Ithaca, N.Y.: Cornell University Press, 1997.

Covell, Charles. *The Redefinition of Conservatism: Politics and Doctrine.* New York: Macmillan, 1986.

Danner, Mark. *Torture and Truth: America, Abu Ghraib, and the War on Terror.* New York: New York Review Books, 2005.

Devigne, Robert. *Recasting Conservatism: Oakeshott, Strauss, and the Response to Postmodernism.* New Haven, Conn.: Yale University Press, 1994.

Fukuyama, Frank. *America at the Crossroads: Democracy, Power, and the Neoconservative Legacy.* New Haven, Conn.: Yale University Press, 2006.

George, Robert P. *The Clash of Orthodoxies: Law, Religion and Morality in Crisis.* Wilmington, Del.: ISI Books, 2001.

Gerencser, Steven Anthony. *The Skeptic's Oakeshott.* New York: St. Martin's Press, 2000.

Gorenberg, Gershom. *The End of Days: Fundamentalism and the Struggle for the Temple Mount.* New York: Oxford University Press, 2000.

Hecht, Jennifer Michael. *Doubt: A History.* New York: HarperCollins Publishers, 2003.

Himmelfarb, Gertrude. *The Roads to Modernity: The British, French, and American Enlightenments.* New York: Alfred A. Knopf, 2004.

Hirschman, Albert O. *The Rhetoric of Reaction.* Cambridge, Mass.: Harvard University Press, 1991.

Hobbes, Thomas. *Leviathan.* First published 1651. Reprinted with an introduction by C. B. Macpherson. New York: Penguin Classics, 1982.

Hume, David. *Dialogues and Natural History of Religion.* Edited by J. C. A. Gaskin. Oxford, Eng.: Oxford University Press, 1993.

Macintyre, Alasdair. *After Virtue.* Notre Dame, Ind.: University of Notre Dame Press, 1981.

Marty, Martin, and R. Scott Appleby, eds. *Fundamentalisms Comprehended.* The Fundamentalism Project. Chicago: University of Chicago Press, 2004.

Mitchell, Stephen. *The Gospel According to Jesus.* New York: Harper-Collins Publishers, 1991.

Montaigne, Michel de. *The Complete Essays.* Trans. Donald Frame. Palo Alto, Calif.: Stanford Press, 1957.

Montesquieu, Charles de. *Montesquieu: The Spirit of the Laws* (Cambridge Texts in the History of Political Thought). Cambridge: Cambridge University Press, 1989. Edited by Anne M. Cohler, Basia Carolyn Miller, Harold Samuel Stone, Raymond Geuss (series editor), Quentin Skinner (series editor).

More, Thomas. *Utopia.* Reprinted with translation and introduction by David Woontton. Indianapolis: Hackett Publishing Company, Inc., 1999.

Murray, Charles. *In Pursuit of Happiness and Good Government.* New York: Simon and Schuster, 1988.

Nichols, Aidan, O.P. *The Theology of Joseph Ratzinger.* T & T Clark, 1988.

Oakeshott, Michael. *Experience and Its Modes.* Cambridge, Eng.: Cambridge University Press, 1933.

———. *On History and Other Essays.* New York: Barnes and Noble Books, 1983.

———. *On Human Conduct.* Oxford, Eng.: Clarendon Press, Oxford. 1975.

———. *Rationalism in Politics and Other Essays.* Indianapolis, Ind.: Liberty Fund, 1991.

———. *The Politics of Faith and the Politics of Scepticism.* Ed. Timothy Fuller. New Haven, Conn.: Yale University Press, 1996.

———. *The Voice of Liberal Learning.* Ed. Timothy Fuller. New Haven, Conn.: Yale University Press, 1989.

Plato. *The Republic.* Trans., with an interpretive essay, by Allan Bloom. New York: Basic Books, 1968.

Santorum, Rick. *It Takes a Family: Conservatism and the Common Good.* Wilmington, Del.: ISI Books, 2005.

Starobinski, Jean. *Montaigne in Motion.* Chicago: University of Chicago Press, 1985.

Strauss, Leo. *Natural Right and History.* Chicago: University of Chicago Press, 1950.

———. *Thoughts on Machiavelli.* University of Chicago Press, 1958.

———. *Liberalism, Ancient and Modern.* Chicago: University of Chicago Press, 1968.

———. *On Tyranny.* Victor Gourevitch and Michael S. Roth, eds. Chicago: University of Chicago Press, 2000.

Sullivan, Andrew. "Intimations Pursued: The Voice of Practice in the Conversation of Michael Oakeshott." Doctoral dissertation, Harvard University, 1990.

Wilson, A. N. *How Can We Know? An Essay on the Christian Religion.* New York: Atheneum, 1985.

Acknowledgments

I owe a debt of gratitude to my agents, Andrew Wylie and Sarah Chalfant, for bearing with me on this often-changing project, and to Tim Duggan, my editor, for persevering with it. The book was also informed throughout by the readers of my blog, www.andrewsullivan.com, who have written countless e-mails challenging arguments developed here and making others. The same can be said for the contributors to the "Beyond Queer" list-serv, who have helped refine and sharpen many of the points in the essay.

Some sections of the book have appeared in print elsewhere before, and have been adapted, abridged, or modified in the text. A rough sketch of my broad argument for the "conservatism of doubt" was published in *The New Republic* in 2005. Parts of the book were also first aired as essays in the *New York Times Magazine*: "The Scolds" from 1998 and "This Is a Religious War" from 2001. My memoir of meeting Michael Oakeshott was first published in *The New Republic* more than a decade ago. My celebration of the "pursuit of happiness" in the final chapter is adapted from an essay I wrote for *Forbes ASAP*. My analysis of the Bush administration's torture policies was developed on my blog and in the *New York Times,* and my scrutiny of the president's signing statements was first aired in an essay in *Time* magazine. My analysis of Oakeshott is also rooted in my own doctoral dissertation, "Intimations Pursued: The Voice of Practice in the Conversation of Michael Oakeshott" (Harvard University, 1990). Other shards of arguments have appeared in differing forms on my blog over the past six years. I have been mulling what conservatism means for the last two decades—and I have tried to integrate those thoughts and pieces of writing into a newly coherent whole.

I'm immensely grateful to those who read drafts or passages of chapters in this book along the way: Scott Horton, Julian Sanchez, Ross Douthat, Steve Groopman, Rob Anderson, Peter Berkowitz, and Chris Grasso. I am indebted to those who first opened my eyes to the world of political theory, especially my dissertation supervisors at Harvard: Harvey C. Mansfield Jr. and the late Judith Shklar. My debt to Michael Oakeshott's body of work is incalculable. My fiancé bore with me through all of this, which is why this book is dedicated to him.

Of course, I bear complete responsibility for any errors, misjudgments, or faults in the book. They are mine and mine alone.

Index